定制办公 4

CUSTOM-OFFICE 4

诠释企业文化与办公空间的融合
INTERPRETATION ON THE INTEGRATION OF CORPORATE CULTURE AND OFFICE SPACE

深圳视界文化传播有限公司 编

中国林业出版社
China Forestry Publishing House

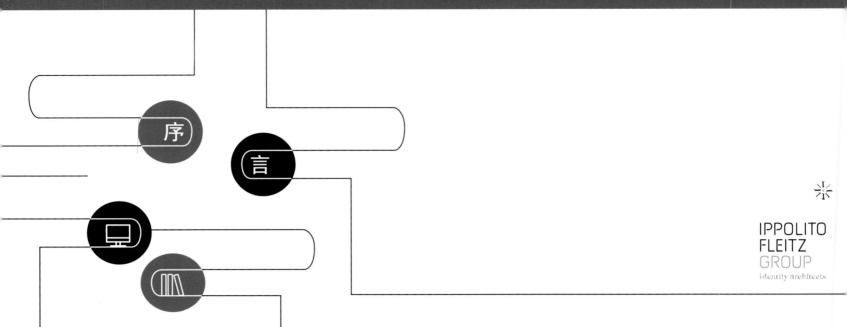

PREFACE

IDENTITY IS THE NEW FACILITY
以个性化为驱动

这已成为一个不可回避的现实：我们所处的是一个深刻数字化和结构性的世界，这要求我们重新组织和设计我们的工作环境。

我们工作的数字化导致所需工具的小型化，这引发了许多变化：工作空间变小，我们的工作也更灵活了。现如今，大多数人的工作都不再局限于办公桌前，我们甚至可以在不同地方工作——办公室、咖啡店、机场等。此外，远程办公和家庭办公室的普遍存在使工作时间变得更加灵活。

一些公司对这类情况采取了应对措施，如取消传统的个人办公桌，转而采用非专属性的办公桌系统。而且，并不是每个员工都在办公室工作，所以相应地减少了办公桌的数量。随着科技产品不断变小，员工也减少了他们所需的物品数量以确保更灵活地工作，所以办公空间的实际需求进一步减少，也因此，空间的使用效率得到了提高。同时，根据每个员工个人的使用习惯，办公桌需以可伸缩的形式来设计，这是一个挑战。

这些都改变了我们对设计办公室的常规认识。一方面，空间的有效利用可以让房地产商产生许多奇妙的反应；另一方面，我们需要开发新的功能区以适合大家工作需求的变化。

新的工作组织意味着新的关联性，是知识与人、计算机、机器和相应环境的跨部门、跨公司甚至是跨工作场所之间透明的连接。创新的跨学科机构，其团队成员皆来自不同的学科领域、部门、地点、机构，或者与机构外的专家合作，跨学科合作旨在集中不同领域的专业知识。这样的合作如果要预约时间，并在同一个地点进行是不容易实现的。

室内布局和设计在这方面扮演着重要角色，能够推动重大的发展。因此，目前的任务不是去拼凑办公桌，而是为了偶然的相遇创造机会。宽敞的茶餐厅、座位围成圆形的空间设计，以及有多个多功能区的大型办公室的扩建，都是目前正在流行的空间设计。

知识工作者总是不断地变化工作内容，其中包括本职工作、客户洽谈会议、创造性解决方案、临时的原型制作、放松身心以及获取灵感等。为了满足这些不同的工作内容和思维方式的要求，设计师寻找着能使他在理想的工作状态中发挥最佳作用的空间。因此，在未来，我们需要减少标准化的工作环境，增加更多样化的工作空间来实现灵活的工作，这些工作室可以将我们带入各自的工作模式当中。

除此之外，工作环境也是企业身份的一部分，它们代表着公司的企业文化，也代表着公司对外界和自身的认同。因为大多数公司都在努力争取市场上最优秀的人才，所以良好的工作环境成为吸引人才的有利条件。

随着工作环境的变化，尊重员工的情感变得比以往任何时候更为重要，因为员工最终的满意度是公司良好形象的表现之一。除了要提供交流和娱乐场所，诸如光、室内气候、声学和自由等因素都是重要的舒适度衡量标准。

室内设计的真正挑战是应对不断变化的工作环境以及满足员工的期望，并把现代工作环境变成塑造身份的地方。

It has become an inescapable reality: the profound digital and the structural change of our world calls for a reinvention of how we organize and design our working environments.

The miniaturization of our digital working tools leads to a multitude of consequences: workstations become smaller, our work more mobile. Today, many of us are no longer bound to our desks. Instead, we can work in different places—in the office space, in a coffee shop, at the airport. Moreover, the ubiquity of telecommuting and the home office has made the workday much more flexible.

Some companies have responded to these changing conditions by doing away with the traditional allocation of individual workstations in favour of a non-territorial desk system. And because not every employee is physically present in the office at all times, a reduced number of workstations will suffice. Actual space requirements are reduced further still as technology continues to shrink in size and employees limit the number of physical things they require to ensure the greater mobility. Spaces are thus employed in a much more efficient way. At the same time, designing the workstations to cater to each user in an individually adaptable and scalable way is an important challenge here.

All this, which changes how we conceive the built environment of work. On the one hand, space can be used more efficiently, which results in many fantastic responses from the real estate side. On the other hand, we need to develop new typologies that suit the changing demands of our work.

Our new work organization means the transparent linking of knowledge with people, computers, machines and the respective environment—across departments, companies or even workplaces. Creative interdisciplinary setups, with team members coming from different disciplines, departments, locations, agencies or working with external specialists are designed to concentrate specialist knowledge from different contexts. This ceases to be easily feasible at one location and on predetermined dates.

In this respect, the interior layout and design play a pivotal role, which can initiate the significant developments. So, instead of thinking about organizing desks, the task now shifts to creating opportunities for chance encounters. Generous tea kitchens, the design of seating niches around circulation areas and the expansion of large offices with multi-functional zones are spatial answers that are currently being employed.

The activity of a knowledge worker varies between focused work, empathic customer meetings, creative solution finding, provisional prototyping, regenerative relaxation as well as reaping the possibilities of inspiration. To meet these different demands on his way of working and thinking, the knowledge worker will look for rooms that put him best in the desired work attitude. So, in future, we need less standardized and more varied environments for the flexible ways of working—rooms that associatively and emotionally transfer us into the respective working mode.

Above and beyond that, working environments are a part of Corporate Identity. They represent a company's culture and identity to the outside and inside world. This is, even more, the case as most companies are fighting hard to acquire the best talents on the market. So the environment becomes a critical asset in the acquisition of human resources.

With such changes to the working environment, it becomes more important than ever to respect the sensitivities of employees. Because ultimately, employee satisfaction is one of the key assets of any company. Offering communication and recreation landscapes is one thing, but factors such as light, indoor climate, acoustics and discretion are also important to comfort criteria.

The real challenge in the interior design is to respond to changing working conditions, to the increased expectations of the employees and to transform modern working environments into places that shape identity.

IPPOLITO FLEITZ GROUP

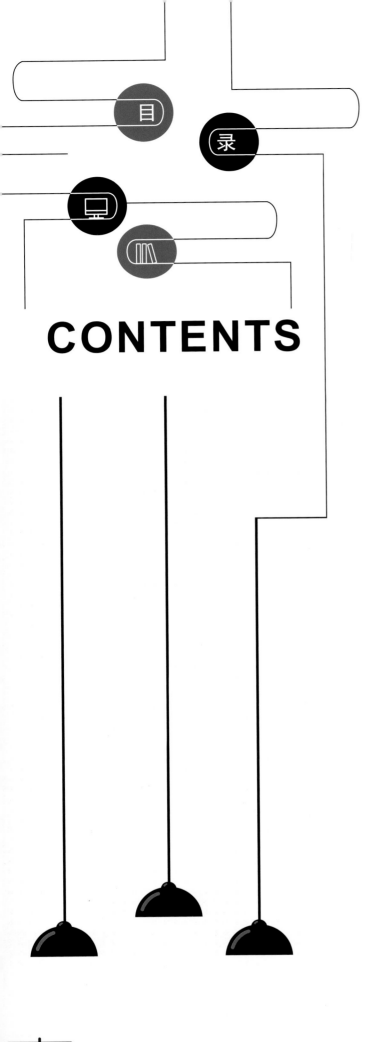

CONTENTS

CREATIVE OFFICE 创意办公

008
Globant Bogota Office
Globant 波哥大办公室

022
Alsea New Headquarters
Alsea 新总部办公室

034
Karma New York Office
Karma 纽约办公室

048
Saatchi & Saatchi New York Office
盛世长城纽约办公室

056
Walking on the Sky
云端上漫步

068
Viola Communications Headquarters
Viola 通信公司总部

080
The Fashionable Office of Modern Finance
现代金融的时尚办公

092
Sberbank of Russia Headquarters
俄罗斯联邦储蓄银行总部

104
MediaCom Warsaw Office
MediaCom 华沙办公室

114
KASIKORN Business-Technology Group Office
KASIKORN 商业科技集团办公室

126
Kolektif House Co-working Space
Kolektif House 共享办公空间

142
Powerlong Ideas Lab
宝龙创想实验室

HUMANE OFFICE 人文办公

162
The Global Advertising Agency INNOCEAN Headquarters
国际广告代理公司 INNOCEAN 总部

170
New · Oriental Aesthetics: Born by Functions and Raise Interesting by People
新·东方美学：因功能而生，因人而有意趣

178
The Pioneer of Shared Work Lifestyle
共享工作生活的先行者

188
Window in the Sky
蓝天之窗

196
Movistar Riders eSports Training Center
Movistar Riders 电子竞技训练中心

204
RD Construction Office
RD 建设办公室

214
The Swiss Branch of the Signa Holding
Signa 控股公司瑞士分部

222
Kochanski Zieba & Partners Law Firm
Kochanski Zieba & Partners 律师事务所

230
Paradox Interactive Headquarters
Paradox Interactive 总部

240
Incanto Moscow Office
Incanto 莫斯科办公室

248
ILGE Bookstore Office
ILGE 的书店办公室

LOW-CARBON OFFICE 低碳办公

260
The Oxygen Bar: Beijing Zhongjun World City FUNWORK
丛林氧吧：北京中骏世界城 FUNWORK

270
King Kungsgatan 36 Office
King 国王街 36 号办公室

282
King Sveavagen 44 Office
King Sveavagen 44 号办公室

298
The Lego Spirit of the Construction Company
建筑公司的乐高精神

310
Porter Davis Office
Porter Davis 办公室

CREATIVE OFFICE 创意办公

- 如果您拥有一个创造梦想的空间
- 可以这样装扮它
- 也可以这样装扮它
- 还可以这样装扮它
- 在这里,你总能找到属于自己梦想的办公空间

GLOBANT BOGOTA OFFICE
Globant波哥大办公室

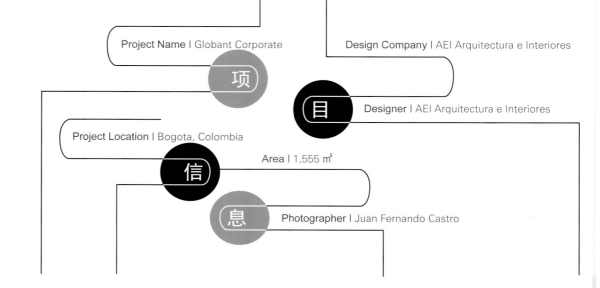

Project Name | Globant Corporate
Design Company | AEI Arquitectura e Interiores
Designer | AEI Arquitectura e Interiores
Project Location | Bogota, Colombia
Area | 1,555 m²
Photographer | Juan Fernando Castro

企业文化 / CORPORATE CULTURE

Globant is a company dedicated to the development of the innovative solutions and technologies, and its brand stands for interaction, experimentation, travel, discovery, playing, incitement, surprise, fun and has unique stories to tell. It was founded in Argentina in 2003. After its trajectory, it has positioned itself as a leader in the Software creation sector in Latin America. Currently, the company has 19 development centers in different cities in South America, with approximately 2,680 professionals.

Globant是一家致力于创新解决方案和技术开发的公司，其品牌代表互动、实验、旅行、发现、玩耍、刺激、惊喜、有趣，并具有独特的故事。该公司于2003年在阿根廷成立，此前它已将自己定位为拉丁美洲软件创作领域的领军企业。目前，公司在南美洲的不同城市拥有19个开发中心，拥有约2680名专业人员。

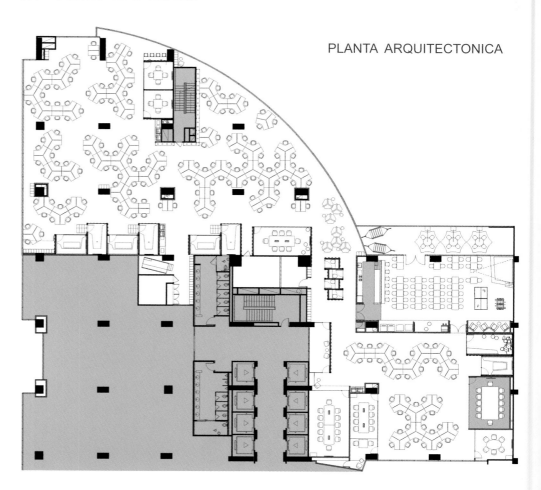

PLANTA ARQUITECTONICA

 设计理念
DESIGN CONCEPT

All of these elements are worked into the design of this project considering the entire experience behind a "voyage", particularly those made over considerable distances. Dozens of different colors, murals, lighting and materials take the users on an ongoing journey that winds from space to space throughout every level of the offices. In doing so, they see that different cultures and expressions are all united in our desire to share them.

公司"旅行"概念背后的全部经验，尤其是那些在一定距离上制造的经验，所有这些因素都被纳入了这个项目的设计中。数十种不同的颜色、壁画、照明和材料让用户进行了一次持续的旅程，从空间到空间，在办公室的每一层都是如此。在这样做的过程中，客户看到的不同文化和表达方式都与设计师分享它们的愿望相契合。

空间特色
SPACE FEATURES

The project has an axonometry typology for work stations, including three employees per section, innovation spaces that contribute to the creativity and productivity of the collaborators. 6 containers with capacity for 4 people, are recycled and adapted for a new use, and thus manage to project their main concept of "trip", emphasizing in the juvenile, dynamic and fun.

Four phone booths and three staff rooms, which are transformed into the conventional rooms by their folding material, are the highlights of the project. The cafeteria is the "chill out area" with a capacity for 60 people. On a wood-like vinyl floor, there are sofas, hammocks, dining rooms and lighting with different designs, the terrace and the playground.

The LED lighting of these offices represents elegance and modernity. By means of hanging lamps of cylindrical and round figures with silver finishes, matt black and gray, the linear hexagon-shaped luminaire occupies the majority of the ceiling that covers the work stations open and signs that guide users in the direction of the different spaces.

这个项目有轴测型的工作站，每个工作站可容纳三名员工，是一个有助于协助作者创造力和生产力的创新空间。6个分别可容纳4人的集装箱被回收利用并适应于新用途，从而突出Globant公司"旅行"的主要概念，强调年轻、活力和乐趣。

这个项目的亮点是将四个电话亭和三间员工室通过折叠材料改造成传统的房间。自助餐厅是一个可以容纳60人的"休闲区"，而在类似木地板的乙烯基地板上，还设有沙发、吊床和不同设计的灯光，以及露台和游乐场。在这部分空间里，员工可以通过不同色彩和不同的功能区，暂时忘却工作烦恼或是其他繁琐，而以放松的状态，在空间中彻底解除平日里的所有束缚，愉悦自己的身心，给思维和灵魂暂时的自由，继而寻找新的灵感，重新投入到工作中。

办公室里的LED照明灯代表了优雅和时尚。亚光黑色的圆柱形悬挂灯和灰色、银色饰面的线性六角形灯占据了工作区大部分开放的天花板，并引导用户前往不同方向的空间。

立面图

ALZADO 1

ALZADO 2

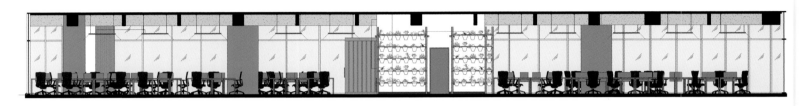

ALZADO 3

ALZADO 4

ALZADO 5

ALZADO 6

ALSEA NEW HEADQUARTERS
Alsea新总部办公室

Project Name | Alsea Brandscaping
Design Company | Cobalto Estudio, Space Arquitectura
Project Location | Mexico City, Mexico
Area | 15,000 m²
Designers | Elisa Ortiz Ambriz, Salvador Fernandez Garcia, Rodrigo Guadarrama Murrieta
Photographer | Rodrigo Guadarrama Murrieta

CORPORATE CULTURE

Alsea is a Mexican multinational corporation specialized in the development of unique culinary experiences for customers throughout Latin America. Their properties include Starbucks, Domino's Pizza, The Cheesecake Factory, Chili's, California Pizza Kitchen, P.F. Chang's and Burger King.

Alsea是墨西哥的一家跨国公司，服务范围涉及整个拉丁美洲，专门为他的客户开发独特的美食经营体验。服务过的客户包括星巴克、达美乐披萨、芝乐坊餐厅、Chili's美式餐厅、加州披萨、P.F. Chang's中餐馆和汉堡王等。

DESIGN CONCEPT

Cobalto Estudio is invited to collaborate in the design and fabrication of custom installations within the new headquarters, specifically for the casual meeting places scattered throughout the five floors. Cobalto Estudio follows a rigorous design process that focused on showcasing the values and core business of each of the brands in Alsea's portfolio. It is very important that they are all represented to ensure that the teams working on each one feel a sense of ownership over the new space.

In addition, all the installations have to comply with strict safety, sustainability and aesthetic standards. To accomplish these goals, Cobalto Estudio works closely both with Space Arquitectura (the architecture firm tasked with the interior architectural design of space as a whole) and with Alsea's internal team in charge of the project.

该公司邀请Cobalto Estudio来设计新总部，并制造室内的定制设施，尤其是为散落在五个楼层的临时会议室定制设备。设计团队遵循严格的设计流程，重点展示Alsea投资的每个品牌的价值与核心业务。重要的是，公司每个部门都派代表参加设计会议，以确保每个小组都能感受到一定的对新环境的拥有权。

此外，所有设施都必须严格要求其安全性、可持续性和一定的美学标准。为了实现这些目标，室内设计团队、负责整个空间室内建筑设计的公司和Alsea负责该项目的内部团队都保持密切合作。

 空间特色
SPACE FEATURES

All the casuals focus on creating areas that are clearly defined, but that do not impede sightlines from throughout space. For the main reception area, Alsea wants a piece that has a large impact, and that makes a statement about who they are and what they do as a corporate entity. Cobalto Estudio creates a wall installation based on laser-cut, powder-coated steel modules with a repeating cutlery pattern that is backlit to create a sense of wonder.

Each of the brands in Alsea's portfolio inspired one or more of the informal meeting areas, one each by Burger King, Starbucks and The Cheesecake Factory, two by Italianni's, and three by Domino's Pizza. However, it was of vital importance that the casual spaces conveyed the values of the brands without being too literal.

所有的休闲区都专注于创建明确的界限，但这些区域不会阻碍整个空间的视线。对于主要的接待区，Alsea希望在这里有一个对公司而言具有代表性的设计，这个设计要让人能够理解Alsea公司的业务和形象。因此设计师创造了一个基于激光切割和粉末涂层钢模块的墙式安装，其中重复的餐具图案可以产生强烈的视觉效果。

Alsea的每一个品牌都是非正式会议区域设计灵感的来源：一间汉堡王会议室、一间星巴克会议室、一间芝乐坊餐厅会议室、两间Italianni's餐厅会议室、三间达美乐披萨会议室。这中间至关重要的是——休闲区所传达的品牌价值都有一定内涵，不会太过于片面化。

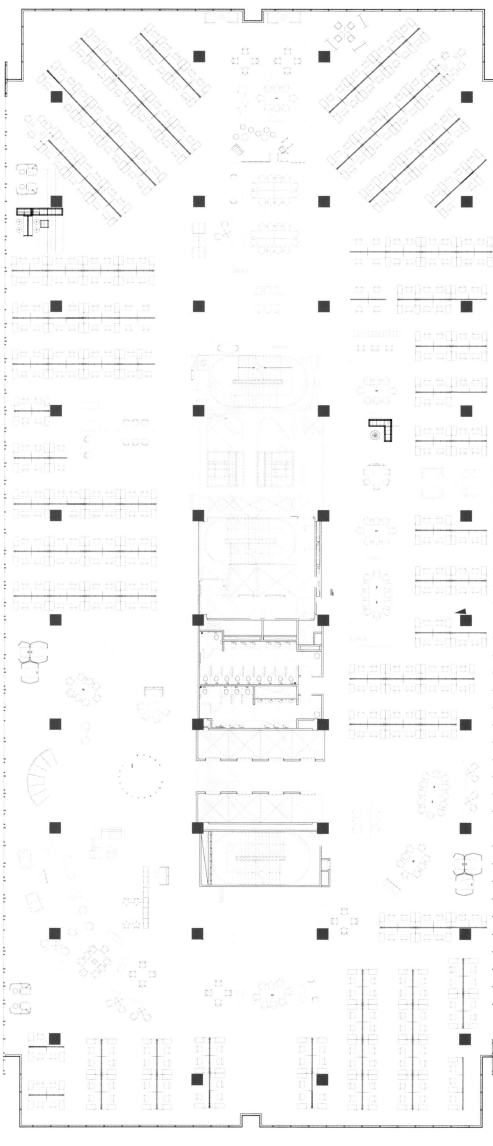

Cobalto Estudio worked to find the essence of these brands and showcase them in a subtler way. For instance, one of the casuals based on the Italianni's brand (a casual Italian eatery) was inspired by pasta drying racks, while the other mimicked flowing spaghetti. Likewise, the casuals for Domino's Pizza reinterpret the iconic logo and utilize the shape to generate a repeating pattern that is pleasing to the eye, identifiable and integrates with the color and material palette in the surrounding area. The casual for The Cheesecake Factory takes the shape of the brand's famous cakes and abstracts it into a flowing pattern of plywood triangles that run from a wall to a hanging ceiling.

设计团队致力于寻找这些品牌的精髓，并以更微妙的方式来展示它们。例如，Italianni's意大利休闲餐厅，其中一间休闲区受到意大利面烘干架的启发，而另一间则模仿流动的意大利面条。同样的，达美乐披萨区的休闲空间重新解释了标志性图案，并利用这个形状生成一个令人赏心悦目的、而且辨识度高的重复图案，并与周围环境的颜色、材料、调色板相结合。芝乐坊餐厅区的休闲空间采用了该品牌著名的蛋糕形状，并将其抽象成一个流动的落地三角形胶合板。

KARMA NEW YORK OFFICE
Karma纽约办公室

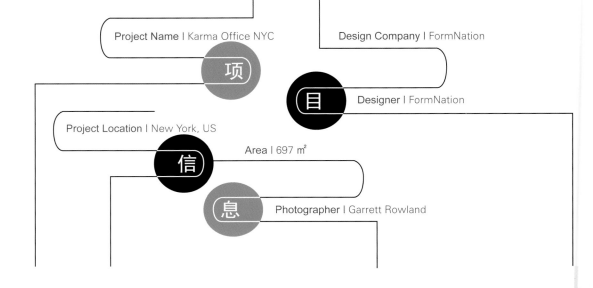

- Project Name | Karma Office NYC
- Design Company | FormNation
- Designer | FormNation
- Project Location | New York, US
- Area | 697 ㎡
- Photographer | Garrett Rowland

企业文化 CORPORATE CULTURE

Karma is a tech start up with strong Dutch influences, which brings a clear, simple, honest and powerful wifi product to the US. Without falling into the typical start up design trap of creating an industrial chic loft design with a slide, FormNation relies on its own Dutch background and design influences to create an aesthetic that focused on a mix of clean Dutch design with American design.

Karma是一家WiFi科技初创公司，颇具荷兰特色，它为美国带来了清晰、简单、诚实和强大的无线网络产品。FormNation设计工作室没有追随当下的工业风设计潮流，而是依靠自己的荷兰背景和设计影响力，创造了一个融合荷兰设计特色与美国设计特色的美妙空间。

设计理念 DESIGN CONCEPT

Located in New York's Little Italy, Karma's office is transformed by the design studio FormNation into a modern space that combines Dutch and American design influences and provides transparency, communication and work-life balances for the employees.

Karma's design brief was short, but clear: design a space with 40 desks, ample break out areas, a war room, communal lunch space for 40, that fits the brand, and they have to move in there in 4 weeks.

FormNation将Karma位于"纽约小意大利"——曼哈顿的办公室改造成了一个现代办公空间。建筑师取荷兰和美国的设计优势，将办公室做成了一个透明的、便于交流的空间，使得员工的工作和生活在这里可以得到平衡。

Karma对设计的要求简单明确：希望办公室的工作区可容纳40张办公桌、一间会议室以及可容纳40人的公共食堂，并且要有宽敞的休息区，整体空间要与品牌特色相符。此外，他们希望可以在四个星期内投入使用。

SPACE FEATURES

The L-shaped space boasts a 14' tin ceiling in an industrial loft with an abundance of the natural light from the large windows along one wall that overlooks Little Italy and provides the perfect environment for collaboration and communication.

Space is divided into 3 main areas:

Work: A colorful pathway of blue and gray floor tiles welcomes employees and guests alike into the office. Employee desks are laid out in a clean grid, comfortable lounge seating exist in the center aisle for conversations and the windows are visible from all desks so everyone can enjoy the view and sunlight. The large, white Delta lights by Rich Brilliant Willing create a cozy work environment and the perfect working light. To top off space is an oversized company logo, in iconic Dutch orange, resembling an all-American gas station.

Meet/private: FormNation built an extension to the existing conference rooms to create an open war room for meetings and

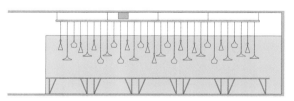

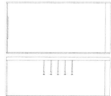

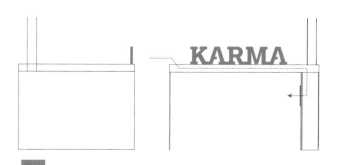

pantone 172C   electricity for tv source on top of meetingrooms

brainstorming sessions in a central location. The conference rooms offer transparency through their glass walls and functionality as the walls are often used to write meeting notes.

Relax/break out: Adjacent to the meeting rooms is a lunch space big enough for the entire office with a long communal table and 40 modern handmade lamps that provide a mix of Brooklyn industrial design and European flair. Following Dutch tradition, Employees often enjoy their coffee in the break out space featuring gray Moroso sofas and iconic Anglepoise lamps in white.

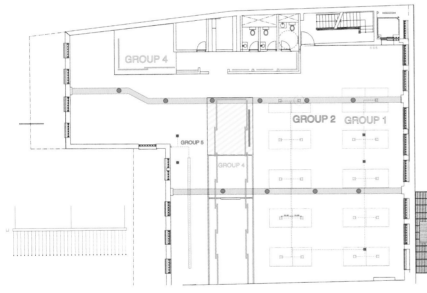

GROUP 1	hanging lamps windowside (9 outlets)
GROUP 2	hanging lamps mid (10 outlets)
GROUP 2	fluorescent light tubes on top of meeting rooms
GROUP 3	fluorescent light tubes on top of meeting rooms
GROUP 4	U-profile with outlets (specifics in attached drawing)
GROUP 5	ceiling mounted lights inside meeting room
⌯	excisting ceiling oulets
▦	fluorescent light tubes (32×4feet,1×2feet)
▭	ceiling light oulets
	light switch
	electrical conduit mounted to ceiling
	extension of meetingrooms

办公室位于一个工业厂房内，整体平面呈L形，14英寸（约35.56cm）厚的锡制天花保证了室内的高度，一侧墙面上的巨大开窗为内部空间提供了充足的自然光照，从窗户可以看到"小意大利"的景观，这为公司提供了舒适的合作和交流环境。

这个办公空间划分为三个主要区域：

首先是工作区：一条蓝灰相间的瓷砖铺成的通道将员工迎入办公区，空间的中部通道上设有沙发，供人休息或交流。办公桌有清晰的网格设置，在任意一张桌子上都可以看到窗外的景色，并享受到自然光照。设计师安置的巨大灯具为整个办公空间提供了舒适的照明环境。公司充满荷兰特色的大型橙色logo，以模仿美国加油站的形式被置于空间上方，将两国的风格融合在了一起。

其次是会议室/私密空间：设计师在原有的会议室外扩展出了一片开放战略商讨区，这个区域位于办公室中央，人们可以在这里进行头脑风暴。会议室的玻璃墙面为这个空间增添了透明感，两侧的墙壁可以用来写会议记录。

最后是休息区：这里与会议室相连，可供整个办公室的人在这里共进午餐。长桌和40盏手工制作的现代灯具使这个空间拥有了两种不同的风格，别具一番风味。设有灰色Moroso沙发和白色Anglepoise经典灯具的休息区可以让他们充分享受自己的午后咖啡。

SAATCHI & SAATCHI NEW YORK OFFICE
盛世长城纽约办公室

Project Name | Saatchi & Saatchi New York Office
Design Company | M Moser Associates
Designers | Charlton Hutton, Jessie Bukewicz
Project Location | New York, US
Area | 4,106 m²
Photographer | Eric Laignel

CORPORATE CULTURE 企业文化

Saatchi & Saatchi (S&S) is a full service, integrated communications network, and they work with 6 of the top 10 and over half of the top 50 global advertisers. They're in the business of getting people to fall in love with their clients' products and services. The company's motto is "Nothing is Impossible", and this concept applies to all our clients and work.

盛世长城是一个全面服务的综合通信网络，他们的合作伙伴包括全球前50名广告商中的半数以上，其中，广告商全球排名前十中有6家公司跟他们合作。公司的座右铭是"一切皆有可能"，他们凭着这一理念对待工作和服务于所有客户。

DESIGN CONCEPT 设计理念

Saatchi & Saatchi (S&S), a global advertisement agency, they engaged M Moser to create an environment that reflects the creative energy inherent within S&S's culture, a space that promotes creative thought and one that is on brand—"Nothing is Impossible".

S&S aspires to have a space where they can easily collaborate in ad-hoc settings, an open environment that allowed for more connectivity amongst teams, and a large communal space where they can conduct town hall meetings and host industry events.

盛世长城是一家全球化的广告公司，它与M Moser设计公司合作，希望创造一个能够反映S&S文化内在创造力的办公环境。这是一个促进创造性思维的空间，也是一个建立在公司信念"一切皆有可能"之上的空间。

盛世长城渴望有一个空间让他们可以轻松地在临时环境中工作，渴望有一个开放的环境允许团队之间更多的交流，还希望有一个大型的公共空间让他们可以举办市政厅会议和行业活动。

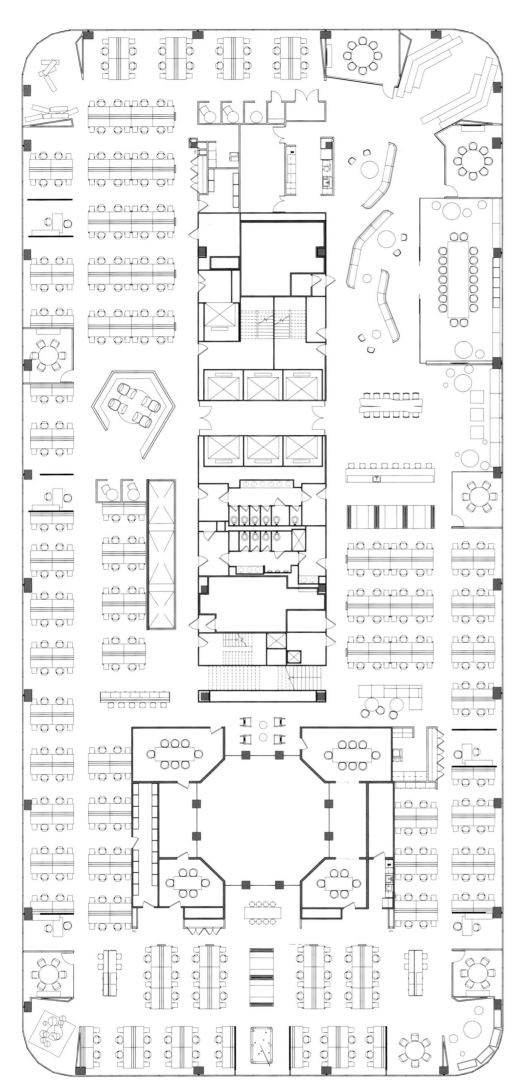

SPACE FEATURES

The 4,106m² floorplate consists of a social hub containing a large cafe bar for communal events, an executive conference center, 220 workstations, informal breakout spaces and an octagonally shaped library. There is also a large 3-floor atrium.

A series of free-form breakout spaces anchor the 4 corners of the floor becoming destination points for Saatchi's design teams to challenge, inspire and rejuvenate the creative mind. Each unique in their function, furniture settings and orientation to the city vistas beyond, they are inspired by different neighborhoods within NYC.

Pencils, a laptop and a Chinese take-out box are some of the many "blacked out" objects mounted to entry wall in the shape of Saatchi's Ampersand (&) logo creating a story of tools woven together capturing the design process of their global advertisements.

这个4106m²的地方有一个社交中心、一个为公共活动设立的大型咖啡厅、一个行政会议中心、220个工作站、非正式的休闲区和一个八角形的图书馆，另外还有一个三层楼的大型中庭。

一系列自由形式的休闲区设定在整个空间的四个角落，这些地方成为激发设计人员创意思维的地方。每一个独特的功能、每一个家具设置、面向城市视野的灵感都来自纽约的不同社区。

安装在入口墙上的有铅笔、笔记本电脑和中餐外卖盒等，这些东西都是公司的标志，设计师创造了一个将工具编织在一起的故事，记录他们在全球范围内广告的设计过程。

← Saatchi & Saatchi 17th Floor

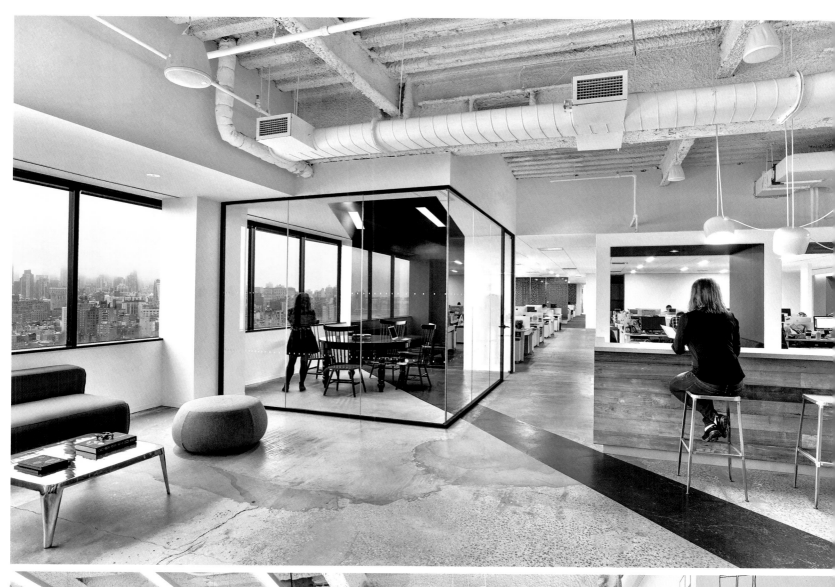

WALKING ON THE SKY
云端上漫步

项目名称 | 广州云硕科技数据中心
设计公司 | Alberto Puchetti—Arboit Ltd
项目地点 | 中国广州
室内设计 | Grace Chu、朱昊恩
项目面积 | 1,800 ㎡
摄影师 | Dennis Lo

企业文化 / CORPORATE CULTURE

Cloud DCS is a data real estate operator focused on creating a cloud ecological base environment. The company's core team members have long been involved in the research, design, construction and operation of data centers. Up to 7 million potential users give the IT industry great potential in China. While the Cloud DCS is the leading company in this industry, and it has always been provided with the network data services for Alibaba, QQ and Baidu.

广州云硕科技是一家专注于打造云生态基础环境的数据地产运营商。公司核心团队成员长期从事数据中心的研究、设计、建设以及运营工作。高达700万的潜在用户让信息科技业在中国拥有极大发展潜力，而设于广州的云硕科技，正是该行业中的龙头大哥，一直为阿里巴巴、QQ和百度搜索引擎等重要客户提供网络数据服务。

设计理念 / DESIGN CONCEPT

The goal of the project briefed by Cloud DCS to the Hong Kong-based design firm Arboit Ltd is to create an identity to the brand in communicating what the company is and how it works. The design of the data centers' architecture and the interior is also an opportunity to interpret the image of the digital science and technology industrial culture and to express its crucial role in today's reality, its history and its future as the industrial core of Cloud DCS.

Using their imagination of the data industry as a starting point, designers transform the "data highway" into a symbolic concept of "data crossing the sky". And taking the image of shuttling through clouds as the basis of visual angle design, designers translate the abstract imagination of network data into a kind of spatial language, emphasize the importance of the data industry promoting the development of the society and changing the mode of the daily life as the main media of the modern information dissemination.

位于香港的Arboit设计公司所接收到的指示是为云硕科技打造一个清晰的品牌形象，必须让人容易理解云硕科技所提供的服务以及背后的理念，也希望借着大楼的建筑和内部设计，重新诠释数码科技产业文化的形象，清楚表达云硕作为云端科技的产业核心对当今时代的重要性、对历史的影响以及对未来的憧憬。

设计师以他对数据产业的想象作为切入点，把"数据的高速公路"转化成"数据穿越天空"这样一个象征性的概念，并取在云层上穿梭的意象作为视角设计的基础，把对网络数据抽象的想象翻译成一种具象化的空间语言，强调数据工业作为现代信息传播的主要媒介对于推动社会发展和改变日常生活模式的重要性。

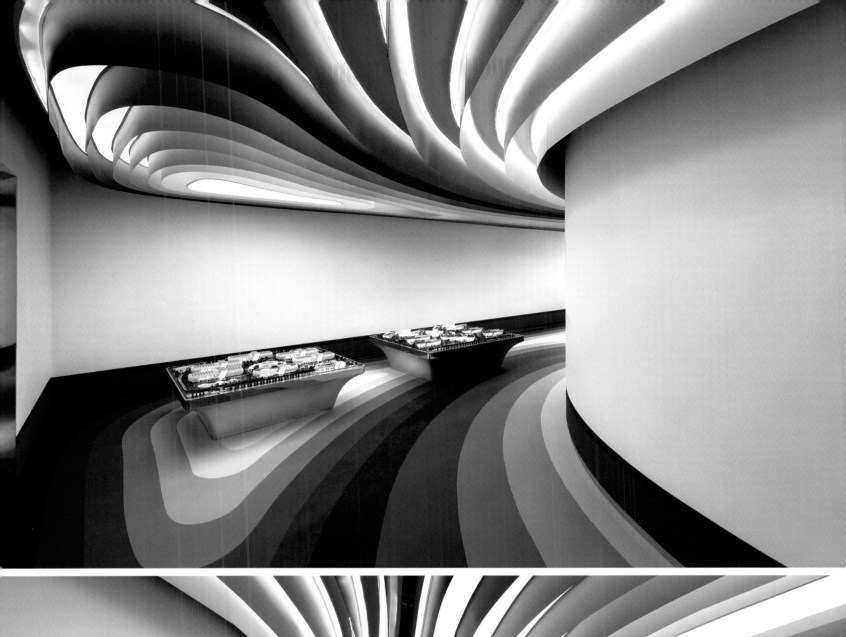
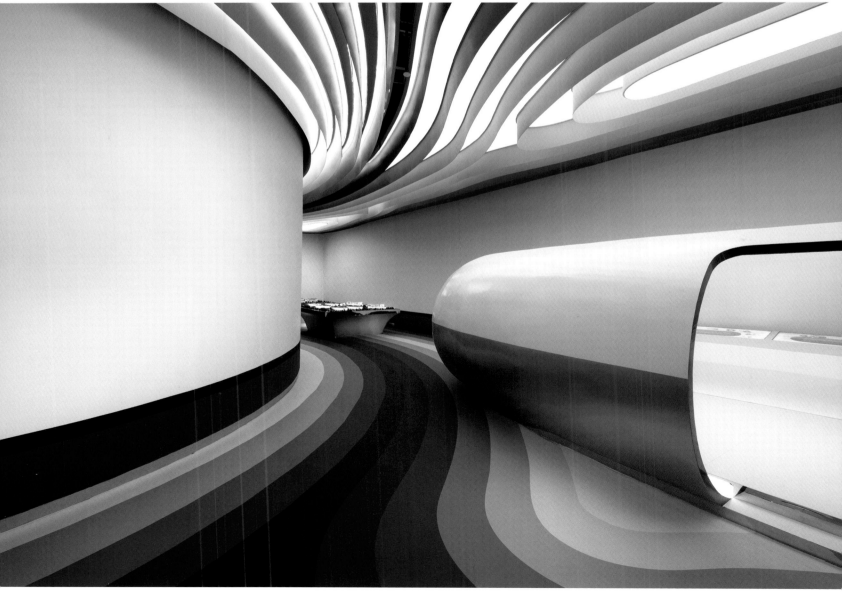

SPACE FEATURES

Taking into account the name of the company and the concept behind the Cloud DCS, the theme of this project is about the image of a stream of clouds containing information carried by winds and flowing in the sky, so that designers use the flowing layers of fading tones of blue to portray today's digital communication. The fluidity and dynamism of clouds are expressed by a 30 meters long sculpture on the ceiling. On the scale and the perspective, its shock power reveals the intangible and powerful properties that transcend the physical and spatial boundaries. The same rarefied artwork defined by the ceiling is reflected on the floor like a gigantic painting printed on the resin floor creating the impression of walking on the sky.

All colors used in the interior design including the signs on the wall, and the decorations are exclusively white and seven tones of blue: these colors define the brand identity. In addition to being functional, most of design features are intended to give visitors an interesting experience that is immaterial, beautiful and dreamy.

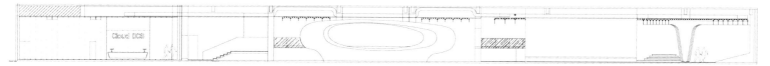

Section BB

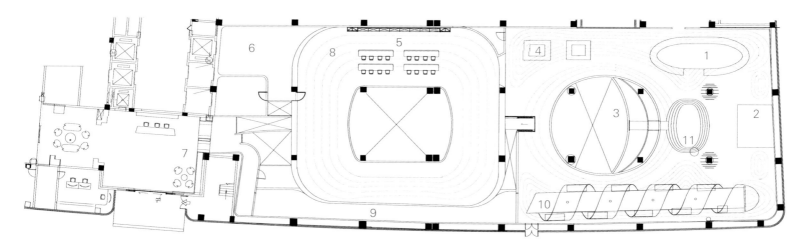

floorplan-ground floor

Inside the data center, a new concept of a factory is proposed: why not developing the intricate lines of the cooling pipes to highlight this unique viewing element between the server racks? The cooling pipes painted different colors are balanced by the digital artwork printed on the wall. The theme of the project is changing the factory from a sad, depressing, productive place to a happy, positive place.

7 Lobby Entrance (160 sqm)
8 Control Room (460 sqm)
9 Projection Tunnel
10 Product Gallery
11 Showroom (660 sqm)

1 Video Pavilion
2 3d Screen Ball
3 Projection Room
4 Model Display
5 Monitor Wall
6 Office

Hallways between the rooms of the data centers are defined by the artworks, the colors and the angle designs, expressing the concept of travelling through the skies. The colorful accessories on the wall allocating with some leisure areas and devices for the break time are aimed to transform a stiff and boring work place into a pleasant working space with the human and functional features. It may be an example of one of the few domestic companies in China that are willing to bravely incorporate such a work-to-entertainment attitude into the office space, and it hopes to break the stereotype that headquarters must be serious and rigid.

考虑到云硕科技的命名和云端科技背后的概念，整个设计旨在表达出数据承载着信息随着气流穿梭于空中的意象，故设计师以层层渐变的蓝色线条来象征这个流动的过程。云层的流动以天花上30m长的造型装置来描绘，规模和视角上的震撼力皆展示了其无形的，能超越物质和空间界限的强大特性。天花上的色彩和线条也一一倒映在地上，以印刷的形式在环氧树脂地板上呈现了一张超大型的壁画，营造出彷佛踏在云层之上的错觉。

整个展览厅的内部设计，包括墙上的标牌和所有摆件的主色皆为白色和七个色调的蓝，这些颜色是品牌形象的主要视觉担当。大部分的造型装置除了功能性以外，也是希望带给参观者一种超越空间和物质，美好而梦幻的有趣体验。

在处理数据中心的内部设计时，设计师提出了另外一个概念：利用冷却管错纵复杂的线条来突显服务器机架间这个独有的视角元素，

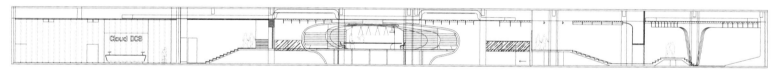

Section AA

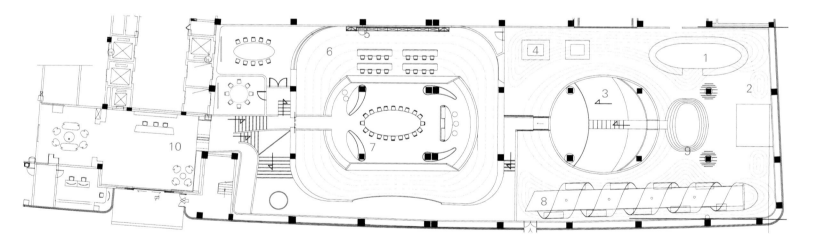

涂上不同颜色的冷却管和印在墙身的数字设计相映成趣，把本来冰冷又苍白，只重视效能的机房变得年轻而有活力。

数据中心内部机组间的走道两旁饰以不同的标牌、颜色和视角设计，同样表达云层上漫游的概念。缤纷的墙饰配合一些传统办公室没有的休憩空间和设备，希望把生硬而无聊的办公室打造成一个兼具人性和功能性的愉快工作空间。

floorplan-first floor

	5 Monitor Wall	1 Video Pavilion
	6 Control Room (460 sqm)	2 3D Screen Ball
9 Showroom	7 Meeting Room	3 Projection Room
10 Lobby Entrance	8 Product Gallery	4 Model Display

VIOLA COMMUNICATIONS HEADQUARTERS
Viola通信公司总部

Project Name | Viola Communications HQ
Design Company | M+N Architecture
Project Location | Abu Dhabi, United Arab Emirates
Designers | Alessandra Barilaro, Lorenzo Zoli, Giulio Asso, Marco Barazzuoli
Area | 1,463 m²
Photographer | Giulio Asso

CORPORATE CULTURE / 企业文化

Viola Communications is headquartered in Abu Dhabi and has other offices in Dubai and Cairo, which is a well-established and fast growing UAE based investment group specializing in providing fully-integrated marketing and communications solutions to national and regional firms. As a long standing, Abu Dhabi based group and a leader in the marketing communications sector in the region, Viola has significantly contributed to Abu Dhabi's communications sector and has established itself as a leading 360 degree integrated communications solutions hub in the region.

Viola通信公司总部位于阿布扎比酋长国，并在迪拜和开罗开设办事处，是一家实力雄厚、发展迅速的阿联酋投资集团，专门为国家和地区公司提供全面整合的营销和沟通解决方案。作为位于阿布扎比的一个长期集团和该地区的营销传播部门的领导者，Viola为阿布扎比的通信部门作出了重大贡献，并已成为该地区领先的全面综合通信解决方案中心。

DESIGN CONCEPT / 设计理念

In this project, the relationships between users are set to be a priority and an accomplishing remarkable success to the challenge of mixing the local taste with the modern interpretation. The classic scheme of the typical workplace is re-invented and adopted, and the traditional arrangement of the office functions into "rooms and corridors" is replaced by a fluid, informal and interactive space. Finally, designers create a vibrant and creative atmosphere, which can reflect Viola Communications' philosophy and capabilities.

该项目优先考虑用户之间的关系，挑战当地品位和现代诠释的融合。典型工作场所的一流方案被重新发现并运用，传统的进入"房间和过道"的办公功能布置被流动的非正式互动空间取代，最终呈现出一个可以反映Viola通信公司理念与能力的充满活力和创造力的环境。

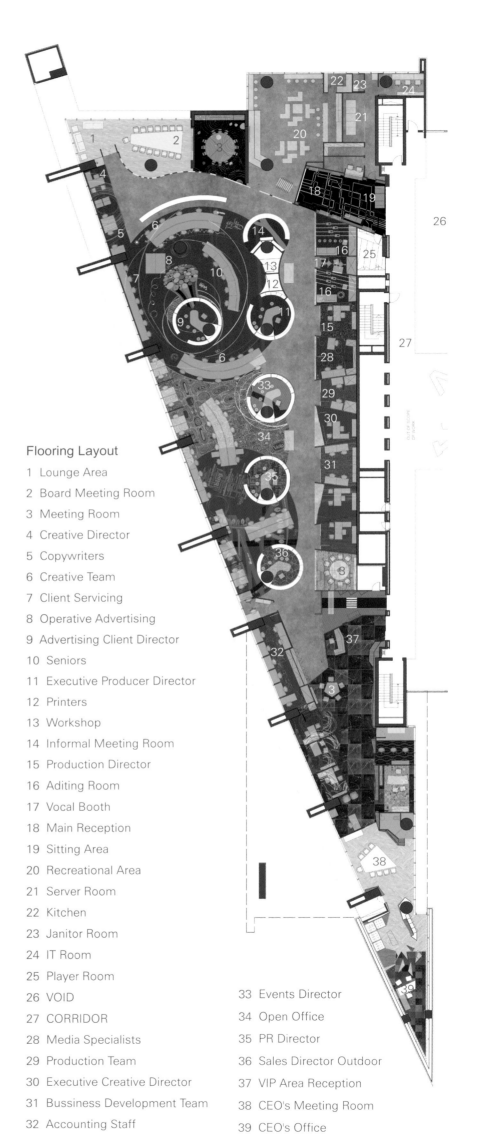

Flooring Layout

1. Lounge Area
2. Board Meeting Room
3. Meeting Room
4. Creative Director
5. Copywriters
6. Creative Team
7. Client Servicing
8. Operative Advertising
9. Advertising Client Director
10. Seniors
11. Executive Producer Director
12. Printers
13. Workshop
14. Informal Meeting Room
15. Production Director
16. Aditing Room
17. Vocal Booth
18. Main Reception
19. Sitting Area
20. Recreational Area
21. Server Room
22. Kitchen
23. Janitor Room
24. IT Room
25. Player Room
26. VOID
27. CORRIDOR
28. Media Specialists
29. Production Team
30. Executive Creative Director
31. Bussiness Development Team
32. Accounting Staff
33. Events Director
34. Open Office
35. PR Director
36. Sales Director Outdoor
37. VIP Area Reception
38. CEO's Meeting Room
39. CEO's Office

SPACE FEATURES

The project is about the new headquarters of the well-known company Viola Communications in twofour54, the media free zone of Abu Dhabi, the capital city of the United Arab Emirates. The design programme is very articulated and includes a mix of open and enclosed offices for around 120 employees (operative staff and directors) from five different business units, as well as three separate entrances and related receptions and waiting areas, a higher management section, several formal and informal meeting rooms, a large recreational area with attached kitchen able to accommodate all employees for the corporate events and gatherings, two video editing studios, one voice recording studio, server room, pantry and storages. There are almost no straight lines in the project, and each piece is kind of unique, as even the themes that look similar or repeated are in fact different from each other in most cases.

The main entrance is inspired by a metro station, where a customized carpet illustrating a colorful metro map connects visitors to introductory screen-boards with engaging presentations of the five departments of Viola, like starting a journey through creativity.

The operative area has a very creative and funky look, definitely something unique for Abu Dhabi. The main office is an open-space, and the zoning is achieved with the use of "the Cylinders" and "the Boxes". Cylinders are open pods containing creative directors' offices, who have their space while being in direct contact with their teams. Boxes contain closed offices, recording studios and meeting rooms fully separated from the main open-space area. All teams can easily interact and work side-by-side while maintaining their environment. Different colors are used to mark each department, and the custom-made design of the carpets also changes accordingly. While the recreational area takes shape from the reinterpretation of the world-famous Tetris. In the second largest meeting room, an irreverent meeting table with swinging chairs brings all meeting attendees back to their childhood, making it very easy to brainstorm with colleagues as well as to break the ice with the most formal clients.

该项目是知名公司Viola通信公司位于阿联酋首都阿布扎比twofour54媒体自由贸易区的新总部办公室。空间里包括可容纳大约120名不同业务部门员工（业务人员和主管）的开放和封闭相结合的办公室、3个独立大门和连带的接待处及休息区、一个高级管理部门、几个正式和非正式的会议室、一个能在公司活动和聚会时容纳所有员工的附带厨房的大型休闲娱乐区、两个视频剪辑工作室、一个录音室、服务器室、食品室和存储室。此外空间里几乎没有直线条，每个区域都是独一无二的，即使不同区域的主题看起来相似，实际上也互不相同。

主入口的设计灵感来自于地铁站，定制的地毯展示了五颜六色的地铁路线图，将参观者带到生动展现Viola五个部门的介绍屏板前，开启一次创意之旅。

工作区有着极富创意的时髦外观，还有某些于阿布扎比而言独特的东西。主办公室是一个开放空间，在分区成功使用了"圆柱"和"箱子"。"圆柱"部分呈开放式，里面有创意总监的办公室，让他们拥有自己空间的同时可以和团队直接联系。"箱子"部分包括封闭的办公室、录音室和会议室，与主要的开放空间完全隔开。所有团队都可以轻松互动、并肩工作，同时保有个人环境。每个部门用不同的颜色进行标识，地毯的定制设计也作相应的改变。而休闲娱乐区是对俄罗斯方块的重新诠释，第二大会议室里带摇摆椅的会议桌将所有与会者带回童年时代，鼓励人打破僵局，集思广益。

THE FASHIONABLE OFFICE OF MODERN FINANCE

现代金融的时尚办公

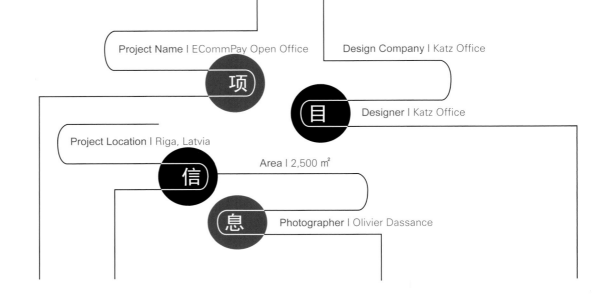

项目信息

- Project Name | ECommPay Open Office
- Design Company | Katz Office
- Designer | Katz Office
- Project Location | Riga, Latvia
- Area | 2,500 m²
- Photographer | Olivier Dassance

CORPORATE CULTURE 企业文化

Headquartered in London, ECommPay employs more than 350 specialists in 7 offices worldwide. ECommPay set out to revolutionize the payment market in 2012, building a fully customisable payment gateway to facilitate transactions via VISA/Mastercard, China UnionPay and popular alternative payment systems. Five years later, our intelligent online payment solution has made us become the partner of choice in helping clients launch into emerging markets, achieve the higher transaction approval rates, lower chargeback ratios and maximize revenues.

ECommPay公司总部设在伦敦，在全球的7个办事处拥有350多名员工。其在2012年着手彻底改变支付市场，建立一个完全可定制的支付网关，以便通过VISA/万事达卡、中国银联和流行的可替代的支付系统进行交易。五年后，ECommPay的智能在线支付解决方案使他们成为帮助客户进入新兴市场、实现更高交易批准率、降低回扣比率的首选合作伙伴，并实现收费比率和收入的最大化。

DESIGN CONCEPT 设计理念

Architectural design studio Katz has implemented a large-scale, unique interior design project on behalf of the international payment solutions provider ECommPay. Each square metre acts as a reflection of the company's corporate culture and core values. At the same time, this office demonstrates how a high-tech interior can embody the corporate values of a modern company, serving as an incentive for the further growth and continued success.

建筑设计工作室Katz为ECommPay国际支付服务设计了一个大规模的、独特的办公室。每一平方米都反映了公司的企业文化和核心价值观。同时，该办公室展示了高科技的室内装饰如何体现出现代公司的企业价值观，从而成为激励公司进一步发展和持续成功的动力。

扫码查看电子版

CREATIVE OFFICE 创意办公

空间特色
SPACE FEATURES

The office interior contains the modern, striking elements and complex geometry: sharp lines, bold patterns, circular panels, hexagonal tiles in a checkered layout. The walls are decorated in an eclectic manner, relying on the shapes of linear geometric figures. The main colors are muted, decorated in calm white, gray and pale blue, and accented by bright pieces of furniture and contrasting panels.

办公室内部包含现代感醒目的元素和复杂的几何结构：锋利的线条、大胆的图案、圆形的面板和格子状布局的六边形瓷砖。墙壁依靠线条和几何图形的形状，以折衷的方式来装饰。主色调是非常柔和的，用平静的白色、灰色和浅蓝色装饰，并用明亮的家具和颜色对比的面板作为亮点。

In all aspects, the office design accentuates the high technological capabilities of its resident company, harmonizing effortlessly with the contemporary styles and the original yet practical approach to organizing working environments. Influenced by the western design, which prioritizes areas for recreation and entertainment above the classic office layout, ECommPay's office is multipurpose. In the open lounge on the 12th floor, for example, weekly corporate meetings, where management discusses corporate news and achievements are held, and educational lectures are broadcast to all global offices. However, the lounge also plays host to cinematic evenings and team celebrations of special events. Employees are welcome to remain after work to socialize, and on Friday evenings, if the week was successful, the company bar is opened. The office building contains an exclusive fitness centre, complete with power and cardio zones, as well as massage and spa facilities.

On each level, the workplace is divided by department and decorated according to the profile of the specialists occupying the area. The manager of each department is positioned in such a way as to have access to each relevant section. The reception is decorated with geometric snow-white panels, complementing the labyrinthine maze stretching from floor to ceiling. The carpet is decorated with white and metallic shades.

Custom furniture introduces notes of modernism while living walls framing the surfaces intertwine delicately with the strict geometric shapes. Some of the desks are equipped with lifting mechanisms, permitting employees to alter the height of their table or to work from a standing position.

办公室设计在各个方面都强调了常驻公司的高科技能力，与现代风格和工作环境的原始又实用的方法相协调。在西方设计的影响下，ECommPay的办公室是多用途的，它优先考虑娱乐区而不是经典的办公室布局。例如位于12层的开放式休息室，每周都会在此召开公司管理层会议，讨论公司动态和成就，并可向全球所有办事处开展教育讲座。休息室也会主持电影之夜或特别的团队庆祝活动。此外，员工们可以在下班后留下来进行社交活动。如果公司在这周取得胜利成果，周五公司酒吧就会开放。办公楼内有一个专门的健身中心，配备了力量训练区和有氧运动区，同时还有按摩和水疗设施。

每层楼上的工作场所都是按部门划分的，并根据该地区的专家概况进行有针对性的装饰。每个部门的经理都能够访问各个相关部门。接待处装饰着几何形白色的面板，与从地板到天花板延伸的迷宫相得益彰。地毯则用白色和金属色调装饰。

定制家具引入现代主义的音符，而生活墙的框架表面上微妙且严格的几何形状交织在一起。有些桌子上装有升降装置，可以让员工调整桌子的高度或站着工作。

SBERBANK OF RUSSIA HEADQUARTERS
俄罗斯联邦储蓄银行总部

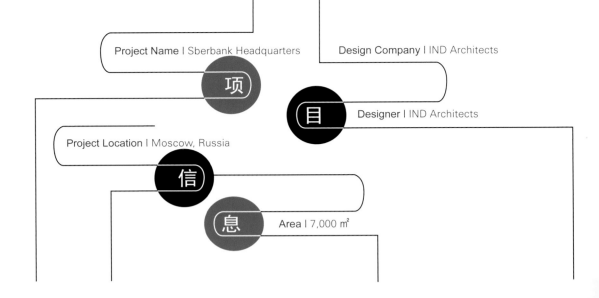

Project Name | Sberbank Headquarters
Design Company | IND Architects
Designer | IND Architects
Project Location | Moscow, Russia
Area | 7,000 m²

CORPORATE CULTURE 企业文化

Sberbank is a state-owned Russian banking and financial services company headquartered in Moscow. The company was known as "Sberbank of Russia" until 2015. Sberbank has operations in several European and post-Soviet countries. As of 2014, it was the largest bank in Russia and Eastern Europe and the third largest in Europe, ranked 33rd in the world and first in central and Eastern Europe in The Banker's Top 1000 World Banks ranking.

俄罗斯联邦储蓄银行是一家总部位于俄罗斯首都莫斯科的金融机构。直到2015年，该公司一直被称为"俄罗斯银行"。该公司在几个欧洲国家和个别前苏联国家都有业务。截至2014年，它是俄罗斯和东欧最大的银行，是欧洲第三大银行，在世界银行排名第33位，在中东欧银行排名第1位，位列世界银行1000强。

DESIGN CONCEPT 设计理念

"The City of Opportunities"—that's how we've named our project, one of the storeys of Sberbank's new office. The concept is inspired by Moscow and its beautiful districts, atmosphere, residents and dynamics that many people of capitals of world's leading countries would envy. The parts of the office are connected by a "ring road". The office itself is divided into six "districts" designed on the basis of landmark sights of a relevant Moscow district: Sokolniki Park, Arbat, Krymskaya Embankment, Gagarin Square, VDNH and Red Square. In addition to the special design, there are meeting rooms in every "district" of the office named after the famous Moscow landmarks, like "Lisya Nora", "Skvorechnik", "Attraktsyon" and other sights of Sokolniki Park. Not only does the design of the office deal with and show its meaningfulness and visual content, but it facilitates navigation for the people visiting it. For instance, the "ring road" helps one easily find their way around in this rather big (7,000m²) office, quicker find a meeting room needed or reach coworkers.

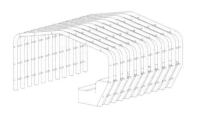
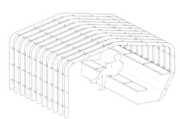

设计师把俄罗斯联邦储蓄银行的新办公室命名为"机遇之城"。这个概念的灵感来源于莫斯科美丽的环境、氛围、居民和动态，而这些也是世界上许多主要国家的首都人民所羡慕的。办公室根据莫斯科相关地区的地标景观而设计：索科尔尼基公园、阿尔巴特街、Krymskaya堤岸、加加林广场、VDNH和红场。整个空间被划分为六个"区域"，各个区域通过一条"环形道路"连接起来。除了特殊的设计外，办公室的每一个"区域"都有会议室，这些会议室以莫斯科著名地标命名，如"Lisya Nora""Skvorechnik""Attraktsyon"以及索科尔尼基公园等其他景点。办公室的设计不仅可以处理与展现公司的意义和视觉内容，并且有利于引导身在其中的人工作或参观。例如，"环形道路"可以帮助人们在这间占地7000m²的办公室轻松地找到自己的方向，也可以更快地找到所需的会议室或同事。

SPACE FEATURES

Sberbank's office is designed in Agile format, and this is literally a huge step up in the country's financial environment. The design provides for all the conditions needed. It streamlines the performance of bank's staff, facilitates communication and hastens the processes. Agile is more than design. It's a special approach to work when cross-functional teams are put together to work on a specific task or product in an effective and prompt manner. Each of six office parts has everything needed for productive work: a coffee point, a reception zone, various kinds of meeting rooms, common and work areas and many other things. Agile work format is good for executives, too. Agile does not suppose large private offices. Executives seat together with staff members in an open-plan area and use all the services in the same way as their teams do. Such changes in the office space idea are likely to make a difference in corporate culture.

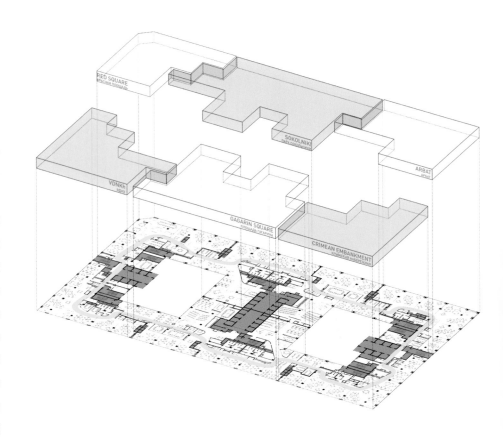

Floorplan-Sberbank

办公室是以方便敏捷的形式设计的，这实际上是该国金融环境的巨大进步。该设计提供了所需的所有条件，简化了银行员工的工作方式，方便了沟通，并加快工作流程。敏捷的不仅仅是设计。当多功能团队被组织起来，以有效和迅速的方式处理特定的任务或产品时，这是一种特殊的工作方法。六个办公室的每个部分都有工作时所需的一切：一个咖啡点、一个接待区、各种会议室、公用的工作区，以及许多其他的东西。敏捷的工作方式对执行人员也有好处，它代表的并不是假设的大型私人办公室。高管与员工一起在一个开放的计划区域工作，并与他们的团队一样使用所有服务。这种办公空间观念的变化可能会改变企业文化。

Floorplan-Sberbank
1 Pillow
2 Frame made of MDF
3 MDF

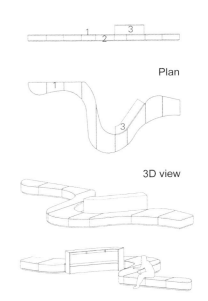

Plan

3D view

Two large areas are arranged in the midst of the office to do presentations, with multi-purpose sofas that can be used to work with laptops as well as stacking chairs and tables. A game zone, a kitchen, the meeting rooms and the leisure zones are arranged here, too. We've provided for various zones: small meeting rooms for quick meets, big meeting rooms for focused discussions, individual rooms for calls, and workspaces with cushioned furniture. Comfortable and diverse furniture adds to concentration and productivity, like high armchairs that enfold you almost all round. Carefully designed details create a warm ambience that encourages effective work and creativity. "Offices in the financial sector are starting to look more like IT companies with their usual atmosphere of certain freedom, creativity and self-expression", says Sberbank's CEO. This is what we were keeping in mind when designing the office for Russia's lead bank.

办公室中有两个大的区域用来做演示,多用途的沙发以及堆叠的椅子、桌子可以配合笔记本电脑一起使用。这里还有一个游戏区、一个厨房、会议室和休闲区。设计师提供了不同的区域:用于快速会议的小型会议室,用于集中讨论的大型会议室,用于呼叫的单独会议室,以及带有缓冲家具的工作区。恰如几乎全方位围绕着的高级扶手椅,舒适多样的家具提高了员工的工作注意力和效率。精心设计的细节创造了温暖的氛围,鼓励有效的工作和创造力。俄罗斯联邦储蓄银行的首席执行官说:"金融部门的办公室开始看起来更像IT公司,他们通常有一定的自由、创造力和自我表达的氛围。"这就是设计师在为俄罗斯的领头银行设计办公室时所牢记的。

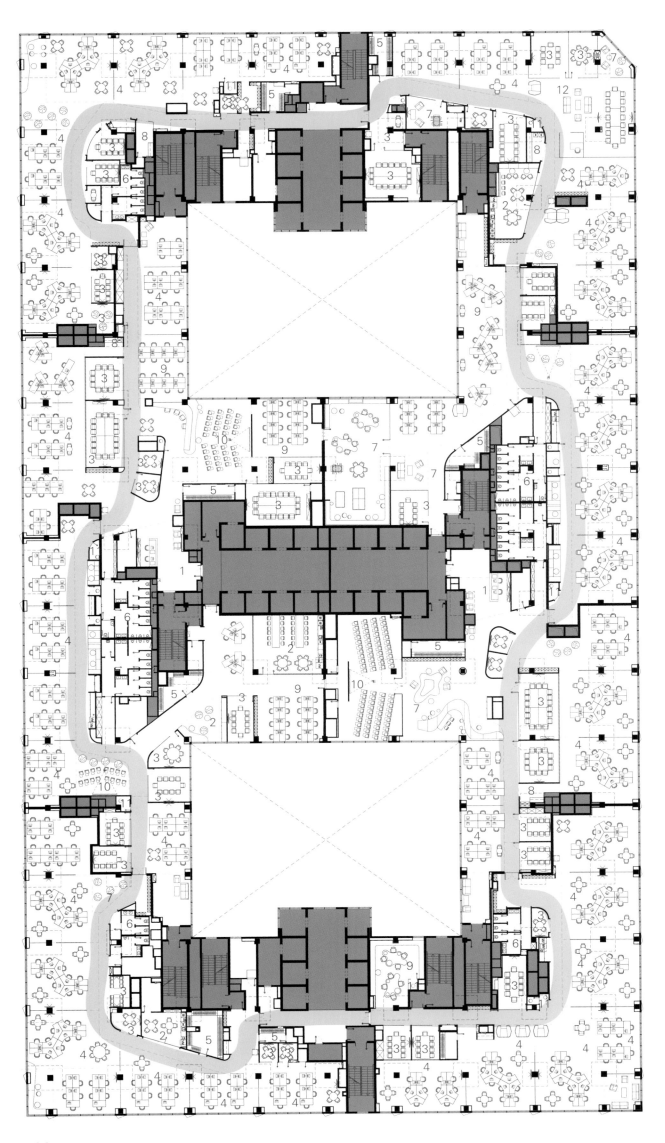

Floor Explication

1 Reception
2 Coffee Point
3 Meeting Room
4 Open Space
5 Wardrobe
6 WC
7 Relax Zone
8 Print Zone
9 Coworking
10 Presentation Room
11 Phone Booth
12 Top Mangement

MEDIACOM WARSAW OFFICE
MediaCom华沙办公室

Project Name | MediaCom Office
Design Company | Workplace Solutions
Designer | Workplace Solutions
Project Location | Warsaw, Poland
Area | 3,000 ㎡
Photographer | Adam Grzesik

CORPORATE CULTURE 企业文化

MediaCom was founded in 1986 in Europe as a subsidiary of Grey Advertising, and soon expanded throughout the world. In 2006, after the acquisition of Grey by WPP, MediaCom became a key part of Group M, WPP's global media operations arm. Group M is WPP's consolidated media investment management operation, serving as the parent company to agencies including Maxus, MediaCom, Mediaedge:cia, Mindshare and Kinetic.

MediaCom作为格雷广告的子公司，于1986年在欧洲成立，它的业务发展很快，并扩展到世界各地。2006年WPP收购格雷之后，MediaCom成为WPP全球媒体业务部门Group M的重要组成部分。Group M是WPP的综合媒体投资管理运营机构，旗下有Maxus、MediaCom、Mediaedge:cia、Mindshare、Kinetic等机构。

DESIGN CONCEPT 设计理念

Based on our research of space usage and work methods in a various organization, we notice that function strengthened with interesting design increase the usage of particular spaces. We decide to use the best developed patterns and inspiration from the urban space and look at the urban planning analogies for the different areas of work in the office.

基于对各种公司及组织的空间利用和工作方式的研究，设计师发现，有趣的设计往往能够使特定的空间发挥出更大的作用。因而设计师决定从城市空间中汲取灵感，并将最前沿的城市设计理念融入到办公室的不同区域。

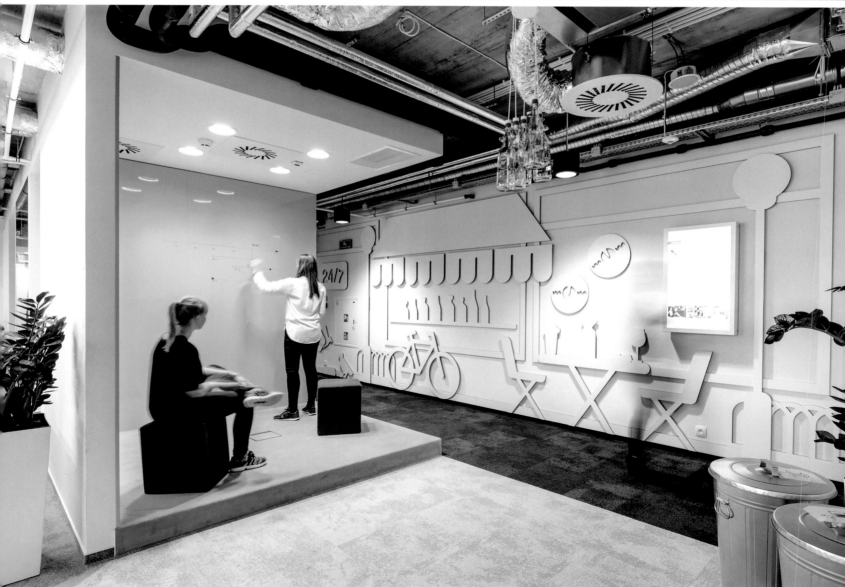

SPACE FEATURES

MediaCom Warsaw that is a part of global holding wants to highlight its local character. The original purpose of the design is to provide a concise and recognizable place to employees and naturally to achieve various functions. From several dozen of Poland capital spots, we choose nineteen that due to their function, unique character and recognition to become the direct inspirations. In this way, designers create the following space: Zbawiciela Square, a wide, open networking zone with kitchen; Warsaw Centralna (Central railway station), the conference spaces dedicated to guests; Kopernik Science Centre, a creative zone; University of Warsaw Library, work in silent/focus zones and meeting places; National Stadium, an auditorium space, perfectly tailored to creative workshop activities, brainstorms etc.

作为全球控股公司的一部分，华沙总部办公室希望彰显其自身的独特性。设计的初衷在于为员工提供简洁、易于辨认的空间，同时能够顺畅地实现各种功能。设计师从波兰的众多城市中选取了19个景点作为功能层面的参照，与众不同的特质和辨识度成为了直接的灵感来源。借此，设计师打造了以下空间——Zbawiciela广场，这是一个宽敞且开放的网络空间，还附带了厨房；Warsaw Centralna（中央车站）是用于接待客人的会面室；Kopernik科学中心是一个创意空间；华沙大学图书馆是一个安静的办公空间兼会议室；国家体育馆则是一间礼堂，是为创意工作和集体讨论特别定制的空间。

The new MediaCom office is located on 3000 square meters and now employs 300 workers, and this flexible design allows to increase it up to 400 people. The office is designed according to the concept of work based on the activity of employees (Activity Based Workplace), providing a wide range of possibilities due to the job activities and expected features. The individual work desk is not assigned to the employee, adding a lot of flexibility to space.

The new MediaCom headquarters along with surprising aesthetic is mainly characterized by its functionality. Communication routes are separated from zones of individual and team work by support zones like meeting rooms, telephone booths or places for ad-hoc meetings. This helps to reduce visual and acoustic distractions generated by people moving around the office and maintains the character of space that is wide open. The natural consequence of our strategy of working space is the introduction of "clean desk policy".

新的MediaCom办公室占地3000m²，目前共有300名员工，灵活的设计使其最多可以容纳400人。办公室的设计以员工活动为依据，为空间赋予了必要的功能和各种可能性。办公桌与员工并非一一对应，这赋予了空间足够的灵活性。

MediaCom新总部惊人地兼具了美感和功能性。会议室、电话亭和临时集会空间等辅助性的功能区域将交流区和工作区分开，减少了人流带来的视听上的干扰，但又保持了空间的开放性。设计策略最终自然而然地实现了"整洁办公桌"的计划。

KASIKORN BUSINESS-TECHNOLOGY GROUP OFFICE
KASIKORN商业科技集团办公室

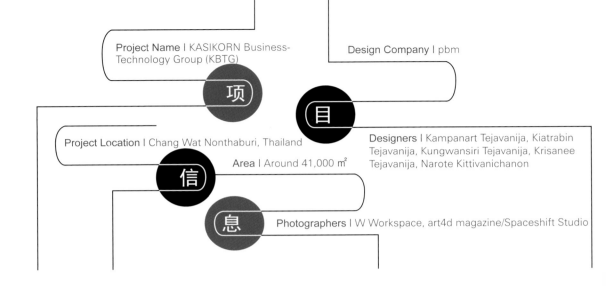

Project Name | KASIKORN Business-Technology Group (KBTG)

Project Location | Chang Wat Nonthaburi, Thailand

Area | Around 41,000 m²

Design Company | pbm

Designers | Kampanart Tejavanija, Kiatrabin Tejavanija, Kungwansiri Tejavanija, Krisanee Tejavanija, Narote Kittivanichanon

Photographers | W Workspace, art4d magazine/Spaceshift Studio

CORPORATE CULTURE

KASIKORN Business-Technology Group is undergoing a major transformation to support the arrival of the new "FinTech" digital banking era. With this industry-wide revolution, KASIKORN Business-Technology Group recognizes the urgency to change its way of working to suit the more modern and seamless culture that is native to its growing tech savvy community. The traditional big, bulky workstations are deemed inhibitive to the communications and mobility across teams and disciplines.

　　KASIKORN商业科技集团正在进行重大转型，以支持新的"金融科技"数字银行时代的到来。面对这场全行业革命，KASIKORN商业科技集团认识到改变其工作方式的迫切需要，以适应日益增长的技术精英社区所需的更现代和无缝的文化。传统的庞大笨重的工作站被视为阻碍团队和学科间的沟通和流动的障碍物。

DESIGN CONCEPT

The new workstation design aims to reduce boundaries and promote the communal exchanges. The addition of creative boxes, recreational area, designated meeting rooms and meeting spaces around office columns combine to create a unique environment which encourages the greater creativity and engagement. While the height of the building in which the office is located and the distance between the expansive row of beams are limited, designers want to make the working area as spacious as possible and must be flexible enough to suit to the changes of future of workstations and accommodate team rotations and restructuring. Therefore, designers pay more attention to the presentation of flexibility and use the mechanical elements to create volume and enhance aesthetics.

　　新的工作站设计旨在减少边界并促进社区交流。创意区域、娱乐区、指定会议室和办公专栏周围的会议空间的增加，创造了一个联合的独特环境，鼓励更多的创造力和参与度。而办公室所在的建筑高度和扩张的梁的距离有限，设计师要让工作区域尽可能宽敞，并且一直具有足够的灵活性，以符合工作站的未来变化，适应团队轮换和重组的需求。由此，设计师更加注重灵活性的呈现，还运用机械元件增加空间的容量和增强空间的美感。

SPACE FEATURES

This office on the 11th floor is the most dynamic space of the entire building. Designers break the height limit of the original building, increase the additional 19 meters of height and make more areas to experiment different design styles.

The main theme is "industrial city" with the interplay of day and night lighting which aims to help the body adjust to the 24 hour working environment. Moreover, K-Stadium is the landmark of KBTG building. Among the thousands of projects that pass through K-Stadium, only the selected ones are chosen to be displayed on this dynamic stage.

Black gives people a strong intuitive feeling, allocating with the bright embellishment color, which is full of texture. The white circular area gathers crowds. Next to it, the green carpet and the yellow single chair on the second floor provide a

comfortable overlooking space. The double-decker individual area involuntarily attracts people's eyes. In addition to the casual brown, calm blue, vibrant green, transparent glass "wall" and the seemingly messy but orderly ceiling, combining all parts together, it is like a small island that gathers into the archipelago, giving people a sense of belonging and the jump thinking.

这个办公室位于建筑的11楼，是整个建筑中最具活力的空间。设计师打破原建筑的高度局限，使得空间增高19m，让其中的更多区域都可以体验到不同的设计风格。

空间主题是"工业城市"，昼夜照明相互作用，旨在帮助身体适应24小时的工作环境。K-Stadium是KBTG建筑的标志性建筑。在K-Stadium的数千个项目中，只有选定的优秀项目才能在这个充满活力的场所中展示。

黑色给人坚毅的直观感受，搭配鲜艳的点缀色，极具质感；白色的环形区间聚拢人群，旁边二层的绿色地毯和黄色单椅提供一个舒适的俯瞰空间；双层的独立区域在不知不觉中吸引人的视线和脚步，此外休闲的棕、沉静的蓝、富于活力的绿、透明玻璃的"墙"、看似杂乱实际有序的天花板，各个部分结合在一起，就像一个个小岛，聚成群岛，带给空间里的人归属感与跳跃思维。

KOLEKTIF HOUSE CO-WORKING SPACE

Kolektif House 共享办公空间

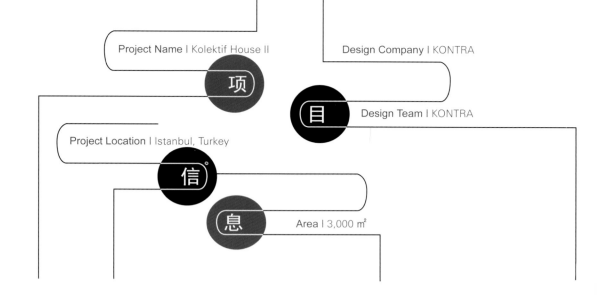

- Project Name | Kolektif House II
- Design Company | KONTRA
- Design Team | KONTRA
- Project Location | Istanbul, Turkey
- Area | 3,000 m²

CORPORATE CULTURE

Kolektif House Levent is an office space rental company, which has thought about everything you would want in your dream workspace and provides the entire infrastructure and service you need, in order to boost your motivation and productivity. Kolektif House Levent is located in Istanbul's financial district, and the design team has turned an old embroidery factory into a living co-working space.

Kolektif House Levent是一家办公空间租赁公司，它能实现你梦想的工作环境，满足你所需的一切，并提供所需的全部基础设施和服务，从而提高成员的工作积极性和公司的生产能力。Kolektif House Levent位于伊斯坦布尔的金融区，设计团队将一家老刺绣工厂变成了一个富有生活气息的工作场所。

DESIGN CONCEPT

Kolektif House II is a continuation of the new generation co-working space concept of Kolektif House Levent. Now, KONTRA has designed the 3rd and 4th storeys of the same building in a different concept, launching the project after a fast-paced period of project and implementation. The purpose of this design is seeking to soften the industrial mood of the project to bring out the identity of its urban character.

The key pivotal points of the interior design of Kolektif House II originate from its focus on the Y-generation. Kolektif House seeks to provide an atmosphere of synergy that generates the highest productivity so that start-ups and both young and mature establishments can nourish and support each other. The common spaces that are designed with meticulous care not only provide a foundation for different kinds of events and meetings but also create a productive network for the companies that gather here under the Kolektif House identity.

Kolektif House II延续了Kolektif House Levent系列的办公空间理念，是其新一代的办公空间概念设计。KONTRA事务所将不同的设计概念用在了同一座建筑的第三、四层空间，经历了快节奏的准备和设计阶段后，该项目顺利面世。其设计目的是充分降低建筑空间的工业氛围，从而为人们带来浓厚的城市气息。

Kolektif House II的室内设计关键因素是对90后这一群体的关注。Kolektif House旨在提供一个具有高效协同生产力的工作空间，以便初创企业与年轻成熟的企业可以在这里相互扶持、相互指教。公共空间通过精心设计，不仅能够提供各种活动和会议的场所，还创建了一个让Kolektif House这个大家庭成员能够欢聚一堂的空间。

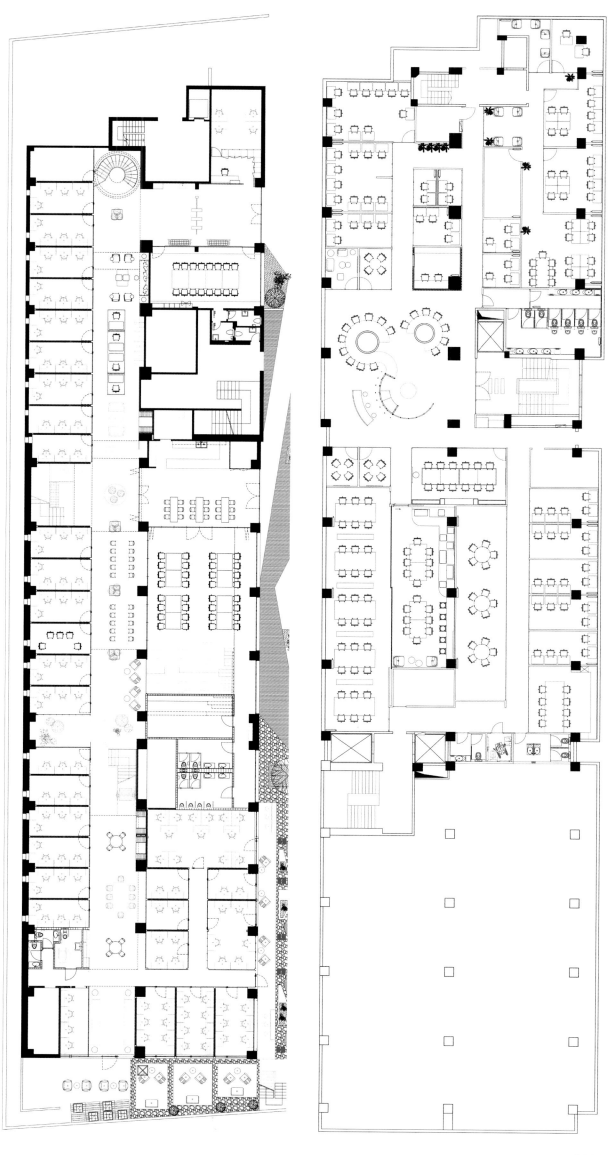

SPACE FEATURES

Kolektif House features semi-open spaces suitable for small meetings, Amsterdam houses, telephone booths and closed personal spaces where users can carry on the private telephone conversations, all of which are only some of the elements of space that add excitement to the common areas. Standing out in their metal structure, the "igloos", with their whiteboard-formatted desks provide users with a practical atmosphere in which to brainstorm. The mini golf links are in keeping with the concept of interior design that was designed to offer staff members the opportunity to break away from the busy workday to engage in an entertaining activity. The self-service "community kitchens" on the floors offer users tea and coffee, microwave ovens to heat meals in as well as a special bar section that can be used to work in. Starting off with the concept of a green office atmosphere in common spaces, the landscaping has been integrated into parts with office furniture to produce surprising results.

The infrastructure of each office space is flexibly designed to meet different needs, featuring plasterboard separators that when removed, offer the opportunity to use different combinations of furniture layouts to achieve the appropriate workplace ambiance. The desks are designed with a functional approach to providing the maximum productivity, featuring accessories such as pencil holders, cable ducts and magazine stands that have been specifically designed and put together for an integrated look.

While the service pipes have been left out in the open as a reminder of the industrial atmosphere of the offices, the design of the lighting apparatus installed with originality inside the cassette ceiling adds a dynamic flair.

Kolektif House有适合小型会议的半开放空间、阿姆斯特丹风格空间、电话亭和封闭的个人空间，员工可以在这里进行私人电话，所有这些都只是一些强调公共空间的元素。该项目运用了钢结构制作而成的圆顶建筑，与白板圆桌一起为员工提供了一个能够集思广益的实用空间。迷你高尔夫设备符合其室内设计的理念，旨在为工作人员提供能够摆脱繁忙公务的娱乐活动。自助式"社区厨房"为员工提供茶和咖啡，还有微波炉能够加热膳食，这里甚至还有一个可以用来工作的特殊酒吧区。从公共空间中的绿色办公氛围出发，将园林绿化部分与办公家具结合在一起，产生令人赞叹不已的艺术效果。

每一个灵活设计的办公空间都能满足不同的需求，当员工取掉划分空间的石膏板隔墙，就能够将家具随意组合从而更好地工作。该办公桌的设计方法十分实用，这样能够为员工带来最大的工作效率，这里还有许许多多的工作配件，如铅笔架、电线收纳器和杂志柜，这些杂物都通过设计而放在一起，形成一个完整的外观。

各种管道设施被裸露在室外，成为办公空间工业氛围的象征，设计师同时也创造性地将照明设备安装在具有磁性的天花板上，这样的手法具有独创性。

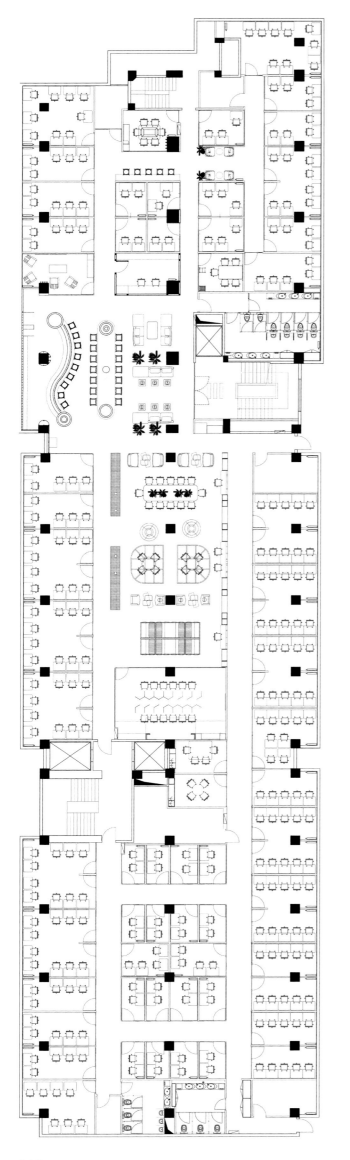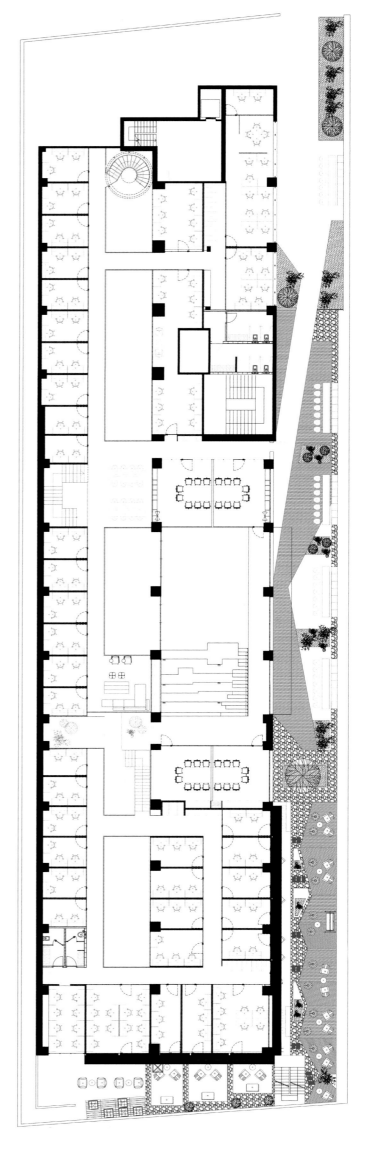

POWERLONG IDEAS LAB
宝龙创想实验室

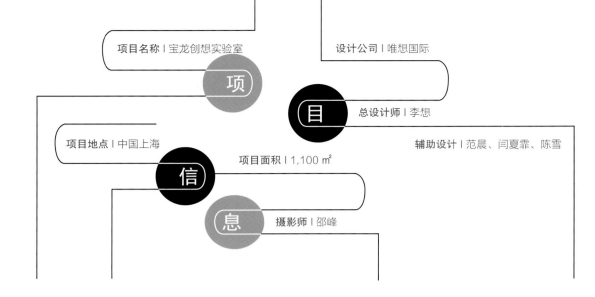

项目名称 | 宝龙创想实验室
设计公司 | 唯想国际
总设计师 | 李想
项目地点 | 中国上海
项目面积 | 1,100 ㎡
辅助设计 | 范晨、闫夏霏、陈雪
摄影师 | 邵峰

CORPORATE CULTURE
企业文化

"Powerlong Ideas Lab" was founded by the Powerlong Real Estate through the integration of various resources and in accordance with the leading Internet research and technology accumulation in the commercial real estate industry and the integration of various resources. It aims to reserve the outstanding new technology and the new species and create the new commercial cornerstone of Powerlong by practicing the new business patterns and the new models.

"创想实验室"是宝龙地产根据前期在商业地产行业领先的互联网研究和技术积累，整合多方资源而成立的，旨在以实践验证新业态和新模式，储备优秀新技术和新物种，打造宝龙的新商业基石。

DESIGN CONCEPT
设计理念

In the 21th century, with the emergence and the gradual popularization of computers, the invisible information age is evolving and changing the humanities and business models of the physical world. Powerlong Ideas Lab is designed to create more laboratory of learning and creating information resources and communication for the society in the information age, and it also hopes that can complete the commercial value hidden behind this era that the consumers and the creators of the information technology can make the research and explore together.

21世纪，计算机的出现和普及让看不见的信息时代推演着、改变着物理世界的人文与商业模式。宝龙创想实验室正是希望能在信息时代的背景下，为这个社会创造更多实验室，供大家创造、学习并传播信息资源，并且希望通过一个复合功能的空间来完成消费者与信息技术创作者对信息技术的共同研发，共同探究这个时代背后隐藏着的商业价值力量。

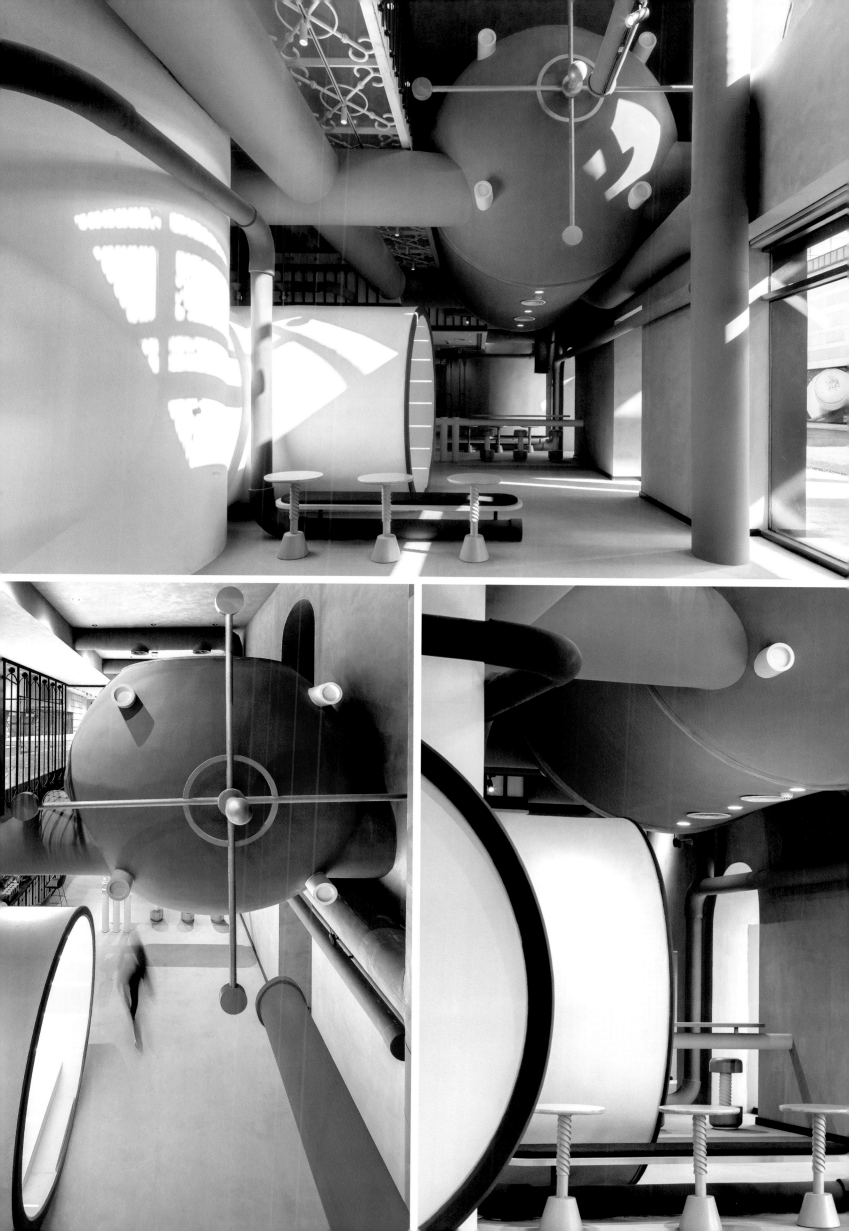

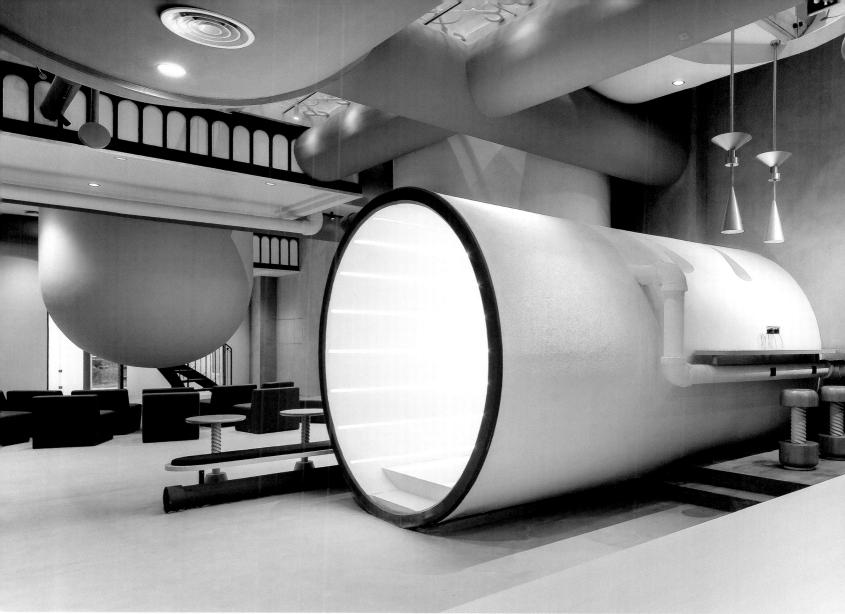

SPACE FEATURES

There is no symbol in the information age, and the information age is based on the quantity and efficiency of the information as the first perception. Therefore, in the design of creating a laboratory, our designers want to borrow the symbol of the previous era as the support and reflection. Because these two eras not only have the ability of disruptive technology as well as research and development but also play the important role in the development of the business.

In the modeling of the whole space, although the designers use the factory in the background of the steam age as the prototype, they still cut out the complicated interlaced parts and only keep some of the most basic functions in the factory, such as the reactor, the energy transmission pipeline and the walking platform for the engineers these kinds of practical compositions. These major functions also reflect some activities that will take place here in the Powerlong Labs.

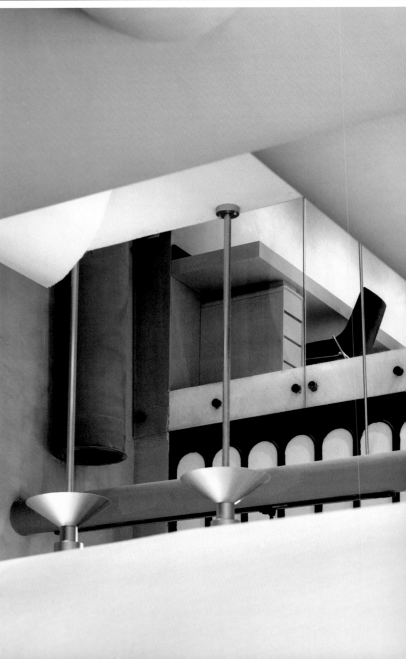

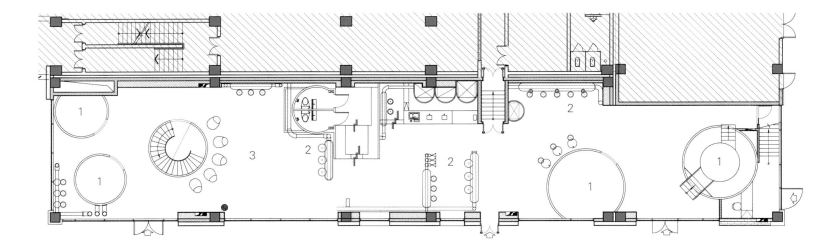

We remove the floor which is originally a two-story space, turn it into an 8-9 meters high space and use the minimalist technique to restore the "reaction tanks" on the ground which will set up the retail experience equipment. Consumers can sense the consumer experience of the new business brought by the different devices in the different cans. And the designers rebuild the floor in the 4-5 meters high space, which is interlaced in the restored "reaction tank" as the work and transportation platform for the research and development staff on the second floor. In this way, the workers work on the platform upstairs, and the consumers experience the business experience from the new technology and information technology downstairs. The eye contact in this height-raised space also allows them to feel each other's existence, making space more interesting.

The whole space is built in the background of a factory composition and takes advantage of the composition of "energy transfer tube" to hide all wires and lines of air conditions which make the exposed industrial design be simplified by the exquisite decorating art. In parts of space, the pipelines are connected to the ground, and a series of tables and chairs for rest are created in combination with the modeling, which makes space look rich in space levels but is also carefully planned for the composition of the different angles of view.

In the selection of the materials, designers prefer to use the clean and cool concrete to depict the skilful temperament of the space form. They also hope to provide a multi-functional and complex office and business model for the research and development in the co-working office and the consumer experiencers who share the office in the same artistic space through the design of the vertical line and sight.

一层平面图

1 体验区
2 无人咖啡区
3 共享会议区
4 洗手间

二层平面图

1 办公区　2 会议区　3 洗手间

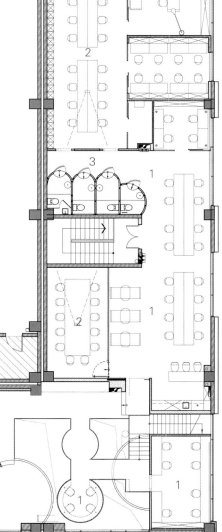

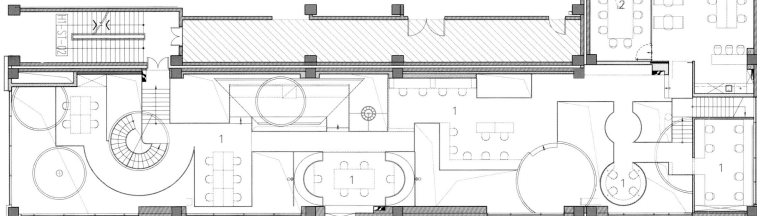

信息时代是没有标志性符号的，它以信息数量与效率作为第一感知，所以在设计这样一间实验室时，设计师更希望能借工业时代的标志性符号作为依托，因为两个时代同样具备富有颠覆性的研发能力与科技产品，同样都对商业的发展有着重要的启发作用。

就整个空间的造型而言，设计师借鉴了蒸汽时代背景下的工厂作为原型，删减了复杂交错的配件，只保留了工厂里最基本的一些功能体，例如反应罐与能量传输管道，还有工程师步行平台等具有实际意义的构图。这些主要的功能体也寓意着在宝龙实验室里要发生的一些相应的活动。

设计团队把原本是两层空间的楼板全部拆除，变成一个高达8~9m的通高空间，地面用极简的手法来还原，这些"反应罐"中会设置新零售的体验设备。消费者可以在不同的罐当中感知不同设备带来的新商业消费体验。在4~5m高的空间中重新搭建楼板，二楼有交错穿插在"反应罐"当中作为共享研发人员工作与交流的平台，这样，工作人员在空中的平台上工作，消费者在楼下的商业体验空间里感受新科技与信息化带来的成果，通高空间的视线联动也使得他们可以互相感知到彼此的存在，让空间变得更加有趣。

整个空间借工厂构图来搭建，在此背景下利用"能量传送管"来隐藏全部电线与空调设备线路，使得空间暴露出来的工业化设计都被精致的修饰艺术简化。在部分空间使管道联动到地面上，结合造型创造了一系列的休息桌椅，使空间看起来层次丰富，同时有构图感。

材料运用上偏向简洁冷静的混凝土，刻画出空间背后的干练气质，并希望通过空间的垂直动线与视线设计为共享办公的研发人员与消费体验者提供一个复合空间，一个可以在同一个艺术空间下办公与商业共存的多功能复合型模型。

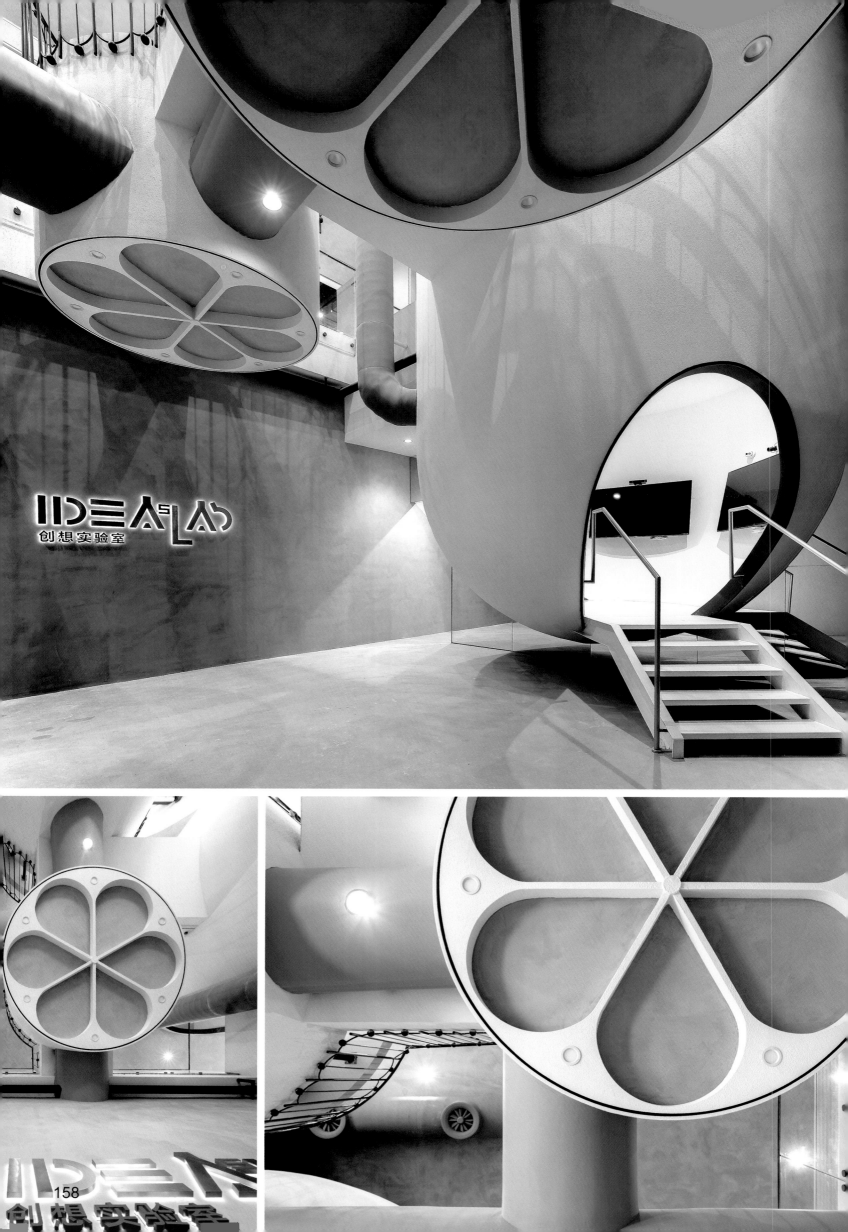

HUMANE OFFICE 人文办公

- 如果您拥有一个创造梦想的空间
- 可以这样装扮它
- 也可以这样装扮它
- 还可以这样装扮它
- 在这里，你总能找到属于自己梦想的办公空间

THE GLOBAL ADVERTISING AGENCY INNOCEAN HEADQUARTERS

国际广告代理公司INNOCEAN总部

Project Name | Innocean Headquarters

Design Company | Ippolito Fleitz Group

Project Location | Frankfurt, Germany

Area | 2,800 m²

Design Team | Andrew Bardzik, Anke Stern, David Schwarz, Frank Peisert, Sebastian Tiedemann, Yuliya Lytyuk, Gunter Fleitz, Peter Ippolito, Daniela Schroder, Tim Lessmann

Photographer | Robert Hoernig

CORPORATE CULTURE 企业文化

Innocean aims to become a global leader by creating future value through customized total marketing strategy service. We understand the true meaning of the things beyond its surface beyond stereotypes, common sense and narrow-mindedness. Beyond the conventional thinking and limitations, we rediscover the new meaning of the world around us.

Innocean的目标是通过定制整体营销战略服务，以创造未来的价值，成为行业中的国际领先者。它主张跳出刻板印象、常识和狭隘的想法，理解事物超越表面的真正意义；突破传统思维和局限，重新发现周围世界的全新意义。

DESIGN CONCEPT

The internationally operating advertising agency Innocean with headquarters in Korea has moved into the new European headquarters in Frankfurt. A flexible and modern work world is created for the young, design-conscious company, which fits the different work zones within the agency.

总部设在韩国的国际运营广告公司Innocean已经搬进了位于法兰克福的欧洲新总部，这个具有设计意识的公司，为年轻人创造了一个灵活而现代的工作世界，适合机构内不同的工作区域。

SPACE FEATURES

Dynamism and movement are the key features of the whole design, which assimilates employees and visitors the second they enter the spacious reception hall. These design elements guide you through the open work zones and the specially created employee library right up to the in-house gym on the fifth floor, which offers an amazing view over Frankfurt. Polygonal spatial elements and a wide range of materials represent the high design standards of the agency itself. The open and transparent work areas, paired with semi-public and completely discreet conference zones promote a creative and communicative working atmosphere.

动力和运动是整个设计的关键特征，当员工和访客进入宽敞的接待大厅时，他们就会不由自主地深入进去。这些设计元素可以引导人们穿过开放的工作区和特别设计的员工图书馆，直接到达位于五楼的室内健身房，在那里可以看到法兰克福令人惊叹的景色。多边形空间元素和丰富的材料代表了机构本身的高设计标准。开放和透明的工作区，加上半公开或完全保密的会议区，这些设计共同营造了具有创造性和方便交流的工作氛围。

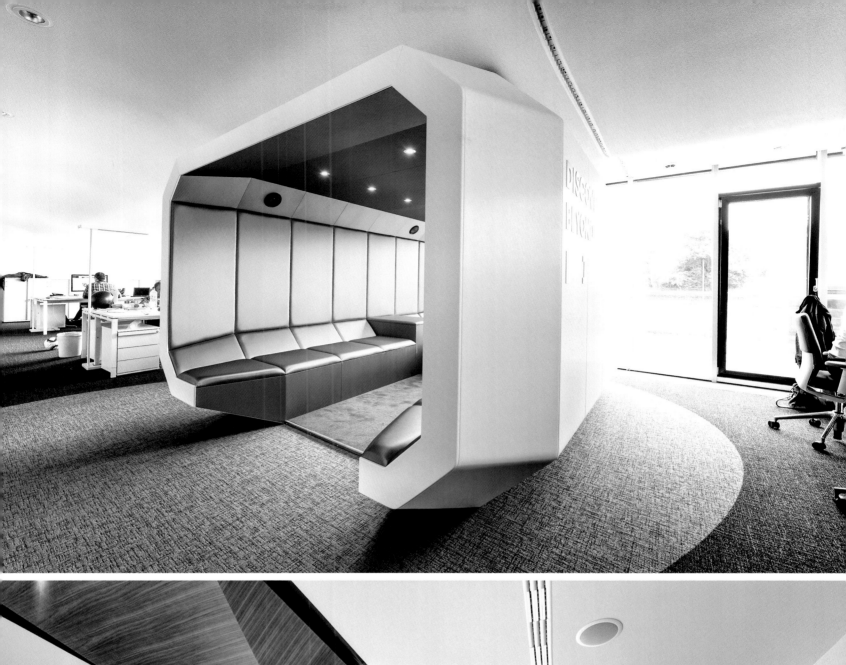

NEW · ORIENTAL AESTHETICS: BORN BY FUNCTIONS AND RAISE INTERESTING BY PEOPLE

新·东方美学：因功能而生，因人而有意趣

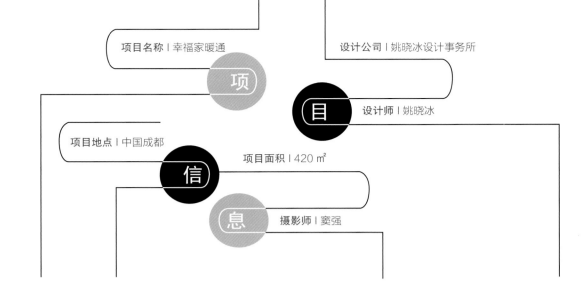

项目名称 | 幸福家暖通
设计公司 | 姚晓冰设计事务所
设计师 | 姚晓冰
项目地点 | 中国成都
项目面积 | 420 ㎡
摄影师 | 窦强

企业文化 / CORPORATE CULTURE

Happy Family·HVAC specializes in providing the services of heating, air conditioning, water treatment and air management systems for the quality-seeking users of families, business and public environment. Taking "achieving a happy life for users" as the mission, the customization to meet individual needs as the standard, the technology as the premise and the extreme as the direction, Happy Family·HVAC is committed to providing users with a healthy, comfortable and pleasant living space environment.

  幸福家暖通公司专注于为追求品质的家庭、商业、公共环境用户提供采暖、空调、水处理和空气治理系统全案服务，其以"成就用户幸福生活"为使命，以满足个性化需求的定制为基准，以技术为前提，以极致为方向，致力于为用户提供健康、舒适、愉悦的生活环境。

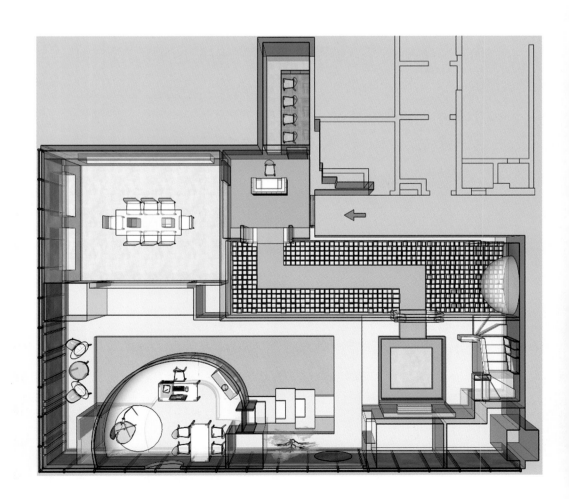

HUMANE OFFICE 人文办公

设计理念
DESIGN CONCEPT

In the reception area, the designer uses the copper element not only to show that copper is the representative of the quality of the industrial products but also to express the enterprise's appeal of the concept of product application, that is buying a good product to get a happy family. The hallway is called "time corridor". Each block represents the memory of the growth of the business, and the ups and downs of the blocks represent the hard experiences of the company's development. Besides, the modelling of the moon is not the end, but the attitude that is the company's pursuit of future.

In the reception area, the poetic description of the landscape makes visitors feel as if they are in the nature to experience the effect that the equipment presents and to show the happy family coming from the healthy air and reassuring water.

Entering into the product exhibition negotiation room, what jumps into eyes is a space combining the modern technology and product technology display together and full of imagination. Space is filled with the description of the future. A round building and a locked expression in the VIP reception room reflect a lifestyle that accentuates the office feelings of the modern executives. In the big conference room, the designer expresses the origami art of "thousand paper cranes" with the modern technique, expresses the demand of functions with the most direct form, advocates the intentional preservation of the original ecological structure of the original building and conveys the harmonious beauty of nature.

在入户前台运用铜板元素，一方面表达铜是工业产品质量的代表，另一方面则表现企业对产品运用理念的诉求，即买放心的产品，得到一个幸福的家。过道被称为"时间走廊"，每一个方块都代表企业成长的记忆，方块的波澜起伏代表企业发展中走过的艰辛历程。另外，月球造型不是终点，而是企业对未来追寻的态度。

在接待大厅，山水诗意的空间描述让到访者仿佛置身于大自然中，体验设备对空间的呈现效果，体现了幸福的空间来自于健康的空气以及放心的水。

步入产品展示洽谈室，一个把现代科技和产品工艺展示结合在一起而充满想象的空间呈现眼前，空间里代入了对未来的描述，充满了前卫的气息。VIP接待室里一个圆形建筑体和一种锁定的表达，其表现的生活方式衬托出现代高管的办公情怀。而在大会议室，设计师用现代手法表现"千纸鹤"折纸艺术，用最直接的形态表达出功能的诉求，主张有意保留原始建筑的原生态结构，传递自然的和谐之美。

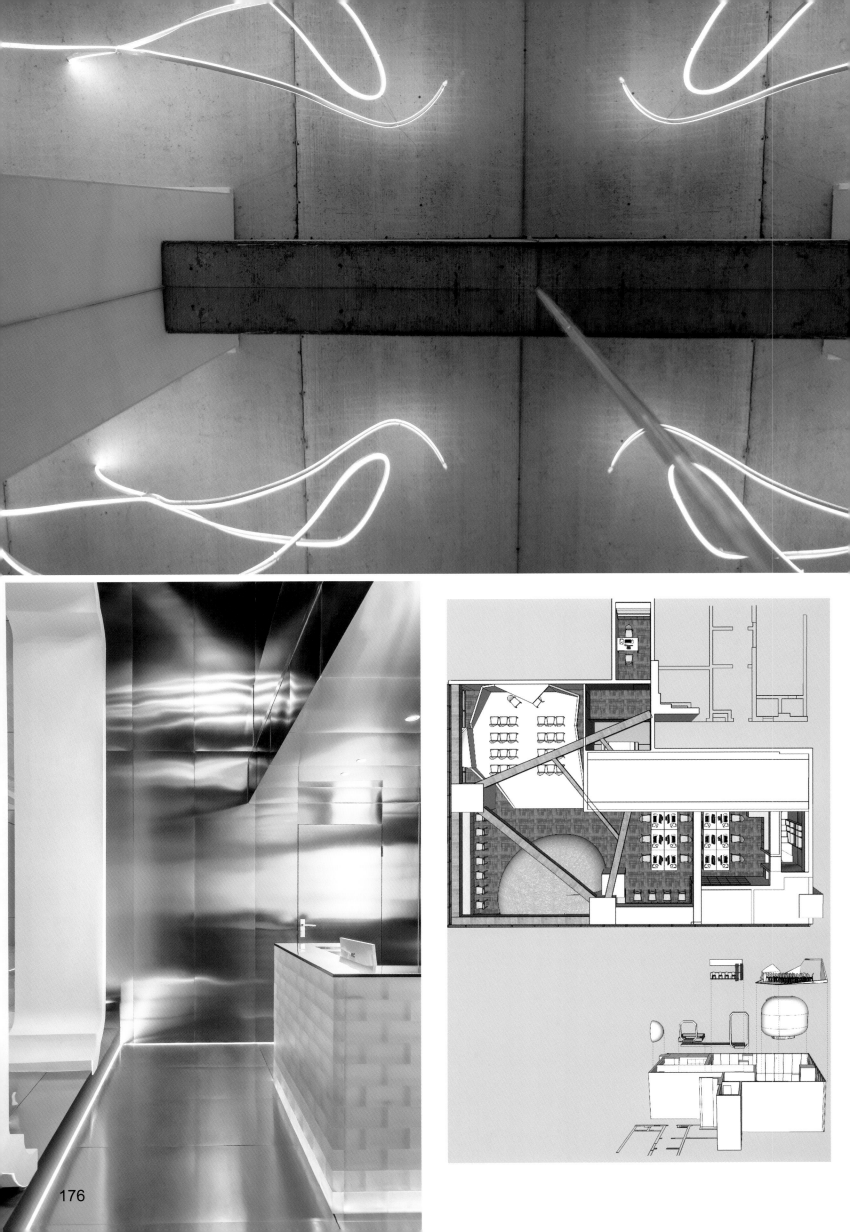

SPACE FEATURES

Each part of space is "separated but not separated, bounded but not bounded". Through the understanding of the Oriental aesthetic culture, the designer uses the concepts of doors, windows, corridors, pavilions, bridges and others to divide a large space into the small spaces with different functions, so that space can have a relationship of a close view and a distant view. At the same time, it does not block the dialogue between spaces, giving space an oriental aesthetic interest.

People will enjoy the different scenery in every step. The designer divides the original 5.8 meters high space into a space containing several floors. Through the division of the functions, every floor has the description of the distant, close, high and low space and time order to make the experience in space orderly and interesting.

With the influence of the natural light and environmental light on space, the white space is like a piece of white paper, which reflects the wind and shadow on it, so that it has the virtual and real change of time and space. It highlights the level, nature and harmony of space and shows the artistic uniqueness of space.

各部分空间"隔而不隔，界而未界"。设计师通过对东方美学文化的理解，运用门、窗、廊、亭、桥等形态理念，将一个大空间分割成不同功能的小空间，让空间产生近景和远景的关系，同时不挡住空间与空间之间的对话视线，赋予空间东方的审美意趣。

人在空间，移步换景。设计师将原本最高5.8m的空间分割成多层空间，通过功能的划分，让每一层都有远、近、高、低的空间顺序和时间顺序的描述，让畅游于空间中的体验变得有序、有趣。

随着自然光线、环境光线对于空间的影响，白色的空间如同一张白纸，映射在其上的光和影，发生时间和空间上的虚实变幻，更加突出空间的层次，自然、和谐，展现出空间的艺术独特性。

THE PIONEER OF SHARED WORK LIFESTYLE
共享工作生活的先行者

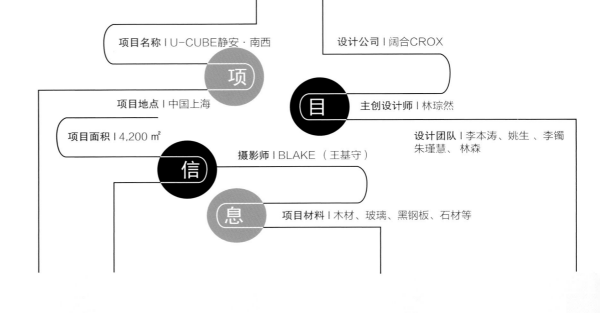

项目名称 | U-CUBE静安·南西
设计公司 | 阔合CROX
项目地点 | 中国上海
主创设计师 | 林琮然
项目面积 | 4,200 ㎡
摄影师 | BLAKE（王基守）
设计团队 | 李本涛、姚生、李镯、朱瑾慧、林森
项目材料 | 木材、玻璃、黑钢板、石材等

CORPORATE CULTURE

U-CUBE is the leading brand of the high-end shared office space in China, uses the "integrating with the creative, sharing the growth" as the concept of the brand and concerns on providing the high-quality entrepreneurial space and humanistic service for the growth-type innovation enterprises.

　　U-CUBE是中国高端共享型办公空间的领先品牌，以"融合创意、共享成长"为品牌理念，专注于为成长型创新企业提供高品质创业空间和人性化专属服务。

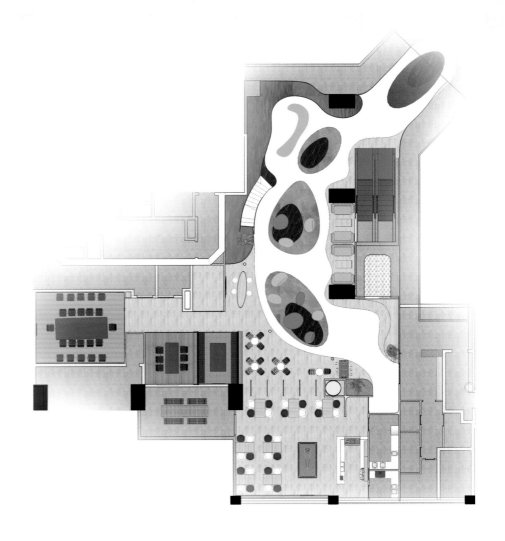

HUMANE OFFICE 人文办公

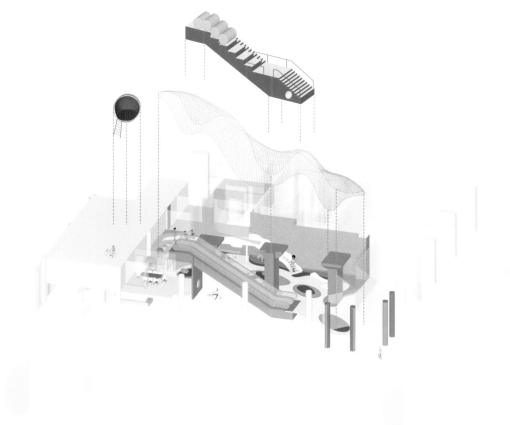

轴侧分析

DESIGN CONCEPT

The world's industrial type goes from the simplification to the complex, and the cross-border collaboration creates a trend of "shared offices" concept. Designers break the boundaries of the traditional offices in response to the industrial diversity and space requirements. Starting from "equipping with sharing, living with nature", we aim to create a space with diversity shared office type. The creative design could redefine the using boundaries of spaces.

世界的产业型态从单一化走向复合式，跨界合作与创客潮流的思维催生了"共享办公室"的概念。为适应产业多样性与空间需求，设计师打破传统办公室的空间界线，以"集合分享，自然生活"为出发点，将坐落于上海静安区的U-CUBE整体空间打造为多样化的共享办公空间，以创新思维重新定义空间的使用界限。

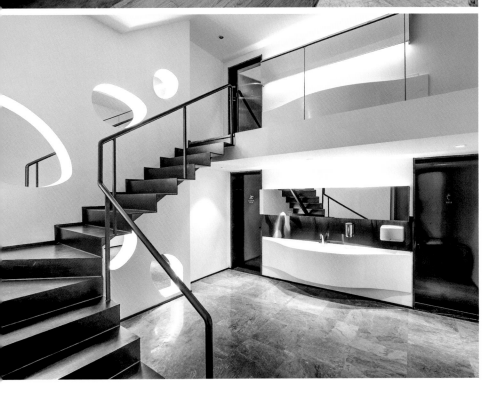

SPACE FEATURES

The Individual Utilization

Through the research, we able to plan the functions of space with the individual's demand, make space meet the public demands and convert them into the practical design elements. For example, we calculate the average amount of the fresh air and the indoor humidity what a person needs, then design the same size green wall and pool to adjust China's dry indoor climate. At noon, some office space converts into a multifunctional restaurant space; at night, it shifts into a bar combining with the hall. Or it can be combined with the flexible partition and hall opening to be a large activity space and provide space for public speeches, salon, exhibition, etc.

A Composite Work Field

U-CUBE contains two floors. Designers decide to add one interlayer to each storey so that a larger space can be made available to fit the appropriate height of an office. It becomes a space that has more layers. Designers also consider different user groups' work demands, so that the meeting rooms in different styles and sizes are arranged and spreads between each floor. Those spaces can be transformed according to the number of people and their mood. The first floor is designed for the big meetings and parties. The second floor is an open space for the easy discussion and communication purpose. The egg shape that floats in the air also plays a positive social role.

The Avant-garde and Natural Idea

Based on the design concept of "Water Dancing", we creatively place a dynamic form into the metropolitan city. This avant-garde, natural design artfully makes the design itself combined with the circulation and dynamic atmosphere created by the U-CUBE. Free cables which are composed of steel tubes will lead you into the interior spaces. They also play a role in connecting the upper and lower floors, changing the ordinary escalators become the focus of the whole creative space. Users can obviously feel its elegance and beauty.

集个别的量化

经过研究，以集个别的概念去规划空间功能，最大可能集约出大众需求，加以转变成可操作的设计元素，如计算平均一个人一天所需要的新鲜空气量与室内湿度，从而在一楼入口放入对等面积的植生墙，搭配调节干燥气候的水池；预设中午办公室用餐最大人数，转换为可多人同时使用的多功能餐厅，而夜间又可将此转变为各界人士共同交流的酒吧，或可配合弹性隔断与大厅开合成一体使用的活动空间，让大厅与此处联动，为公众提供演讲、沙龙及公共展览等功能。

复合的工作场域

CUBE包含两个楼层，而建筑师在一二楼加设了夹层，迎合创意人才对楼中楼的偏好，进而也让空间更有层次。对内考虑不同人群办公的讨论需求，也在各个楼层空间内分布不同大小、不同风格的会议间，让创意在各个角落蔓延，人们可以按照不同心情与人数变换讨论空间。一楼供大会议与聚会使用，二楼强调开放轻松的讨论氛围，让飘浮在空中的蛋形体也扮演积极正面的社交角色。

前卫自然的观点

建筑师以"水舞"为概念，在都市丛林内置入一种创意的动感形式，蔓延出前卫自然的设计，让设计巧妙融入U-CUBE的动线与氛围营造之中。由入口开始指引

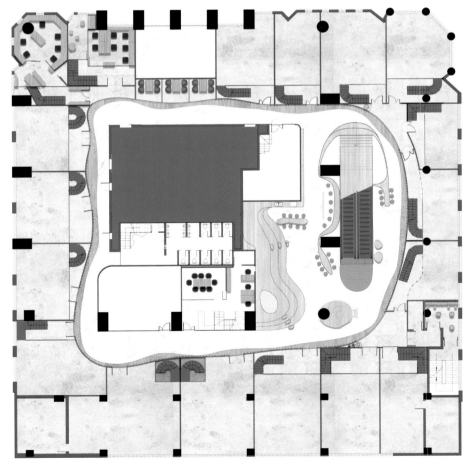

人入内，由钢管构成舞动的自由网线，串接起上下楼层，再普通不过的手扶梯竟也成了整体空间的创意焦点，宛如韵律般的飘逸性是整体给人的感受。

WINDOW IN THE SKY
蓝天之窗

Project Name | Window In the Sky

Design Company | Ippolito Fleitz Group GmbH

Project Location | Frankfurt, Germany

Area | 4,000 m²

Designers | Gunter Fleitz, Hansen Hermawan, Peter Ippolito, Tina Jochens, Kamil Kaczmarek, Christian Kirschenmann, Melanie Neska, Isabel Pohle, Jennifer Schafer, Verena Schiffl, Anke Wankmüller

Photographer | Zooey Braun

DESIGN CONCEPT 设计理念

In a business world characterized by a progressive digitalisation of the work processes and the increased staff mobility, locations and elements of emotion and identity become central to the concept of the modern working environments. Consequently, today's "Activity Based Design" needs to focus on providing employees with the working environments that offer a fitting atmosphere of support for the task at hand. It is a matter of projecting structures, offering incentives for the communication and creating places of the encounter.

工作流程的数字化程度日益提高，员工的移动性逐渐增强，这些发展趋势正在不断影响着我们的工作环境，使我们对情感的归属感和身份的认同感的要求越来越高。本次设计的目标是：结构清晰、提供沟通条件、创造交流机会。"以活动为导向的设计"是当今对工作环境的要求，针对各种不同的工作任务提供不同的有利环境。

SPACE FEATURES 空间特色

In response to the changing structures of the work and communication, we've created two and a half floors of the innovative office space, spread over 4,000 square meters in Frankfurt's "TaunusTurm" building. Designers divide each floor into spatial segments, which in turn are segmented into "Neighborhoods". These "Neighborhoods" are connected to "Market Places", where are the additional communication hubs that create a sense of identity. While the library area with its lush greening presents a place of retreat.

Breath-taking views of Frankfurt welcome arrivals in the entrance area on the 35th floor, where a sculptural staircase and a seemingly weightless ring sculpture emphasize the spacious layout of the two-storey lobby. Floating in open space, the mirror-coated ring highlights the panoramic view and creates a connection to the top floor.

Characteristic nude, teal and blue color schemes reinforce the localization of the specific spatial segments. In the office areas, the combination of foils, covers and glass surfaces with graphic applications creates exciting layering effects, while acoustically effective glass partitions, textile-covered panels, thick carpets and heavy curtains provide a high level of soundproofing. Inside the common rooms, we intensify the harmonic play of colors with polygonal color fields. Here, the vigorous color schemes and the surprising graphic elements emphasize the dynamics and energy of the informal spaces of communication.

Complementing the zoning's rhythm is a variety of ceiling-installations. With his "WELTICHLOTSE" project, the linguistic artist Bruno Nagel created word-ceiling-installations customized for the company's areas of activity. Merging terms from diverse fields and subjects, he develops unique word structures that challenge readers with structurally flexible modes of reading and thinking.

Apart from the open space areas for four employees each, the new premises offer a variety of single, double and corner offices. Height-adjustable desks allow for a flexible work flow, whereas the conference tables in the individual offices create the necessary discretion for intimate conversations. Although consistently designed for desk sharing, the closed offices give rise to a pleasant balance between personal offices and public spaces with a sense of identity. Boasting a panoramic view of the skyline, the corner offices are technically shared by two partners but also available to colleagues during absence periods.

Mould-breaking creative spaces complement the workplaces and allow employees to briefly immerse themselves into inspiring new worlds — from the local "Ebbelwoi room" to a wild jungle area and a blue chamber of marvels,

the creative spaces purposely present perfectly contrary themes. With a variety of flexible and differently-sized spaces, the conference room offers state-of-the-art equipment and a maximum of discretion.

这一总计面积达4000m²的办公空间位于法兰克福的TaunusTurm大厦中，上下分为两层半，为各种沟通和工作环境提供了理想条件。设计师对每个楼层做出了空间划分，并开设了多个交流区，这些交流区就像一个个市中心的广场，为进一步交流提供了具有强烈身份认同感的空间属性。在阅览区，郁郁葱葱的绿植作为装饰的同时也提供了一个放松沉静的场所。

位于大楼三十六层的前台拥有俯瞰法兰克福的视野，这是公司的一个重要景观。一道充满雕塑感的楼梯和看似漂浮的环形雕塑令这个跨越两层的大堂更显宽敞。悬挂在空中充满雕塑感的反光环将目光引向室外的景致，并起到上下贯通的效果。

设计师用不同色彩定位不同空间，主要包括裸色、藏青和蓝色。在办公区，贴膜、包布和玻璃表面上粘贴的图案构成跳跃的层次感。玻璃隔断、织物屏风、厚重的地毯和窗帘都能起到良好的消音作用。在公共区，我们将色调调高，多边形色块深浅不一，颜色各异。强烈的色彩和不规则的图形有效地促进了非正式的沟通。

空间的区域划分同时体现在天花板的结构设计上。语言艺术家布鲁诺·纳格尔以"WELTICHLOTSE"为题，把不同领域和学科的术语结合起来，发展出独特的词汇结构，在天花板上进行文字创作，使人在穿梭之间获得各种不同的视角和阅读体验。

除去四人一组的开放式工作区之外，还有其他种类的办公室，如转角、单人、双人等办公间。有升降功能的办公桌使人们可站可坐，方便灵活。单人办公室自带的会议桌使私密谈话有了合适去处。封闭式的办公室同时也支持共享办公；个人封闭的办公环境与公共空间形成平衡，又有一定的可识别度。转角办公室的窗外是法兰克福的天际线，原则上这里是公司两位合伙人的办公室，但二人不在时也允许其他员工使用。

公司里除办公区之外，还辟有创意区。一处是充满故乡情调的"苹果酒屋"，一处是"热带丛林"，还有一处是能令人灵感迸发的"奇迹角落"。无论空间大小和用途，均配有一流的装备并具备极好的保密性。在这里没有条条框框，可以天马行空，放松头脑。

MOVISTAR RIDERS ESPORTS TRAINING CENTER
Movistar Riders 电子竞技训练中心

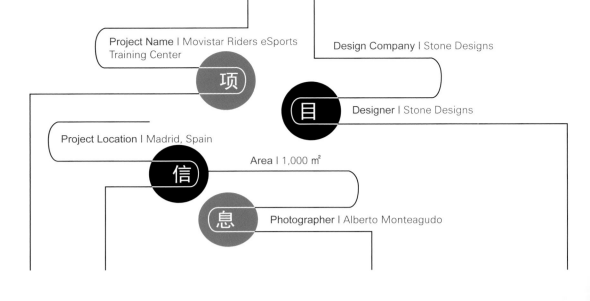

- **Project Name** | Movistar Riders eSports Training Center
- **Design Company** | Stone Designs
- **Designer** | Stone Designs
- **Project Location** | Madrid, Spain
- **Area** | 1,000 m²
- **Photographer** | Alberto Monteagudo

CORPORATE CULTURE

Movistar Riders is a Spanish eSports organization founded in January 2017, and has teams in Call of Duty, League of Legends, Counter Strike: Global Offensive, Overwatch and Hearthstone.

The Movistar eSports Center is the first high performance center in the Europe of eSports and headquarters of the Movistar Riders club. It is a 1,000m² warehouse open to the community and where all the activity of the Club is carried out, as well as the celebration and retransmission of competitions related to the dissemination of electronic sports. There are 4 training rooms and an area with capacity for 70 spectators. In short, a "sports city" born with the purpose of becoming a benchmark and forum for the industry of the videogames and eSports sector.

Movistar Riders是一个西班牙电子竞技组织，成立于2017年1月，拥有"使命召唤""英雄联盟""反恐精英：全球攻势""守望先锋"和"炉石传说"等团队。

Movistar电子竞技中心是欧洲电子竞技和Movistar Riders俱乐部总部的首个高性能中心。这是一个向社区开放的1000㎡的仓库，人们在此开展俱乐部的所有活动，以及庆祝和转播与电子竞技传播有关的赛事。这里还有4个培训室，以及一个可容纳70名观众的竞技场。简而言之，这个"体育城市"诞生的目的是成为视频游戏和电子体育产业的基准和论坛。

DESIGN CONCEPT

Instead of getting inspired by the esthetic of the gaming world, we think that it is much more important to get inspired by the team. They are the soul of this project, and that is why space has to be dedicated to them. They are going to spend a very long time in this space, and our duty is to make them feel at home. And that is what are looking for when we select the materials and the esthetic of the place.

与其从游戏世界的审美中得到灵感，设计师认为更重要的是要从团队中获得灵感。团队是这个项目的灵魂，这就是设计师为什么要将这个空间献给他们的原因。他们会在这个办公室里待很长时间，设计师的责任是让他们有家的感觉。这也是设计师在为这个空间选择材料和创造美学时所要寻找的东西。

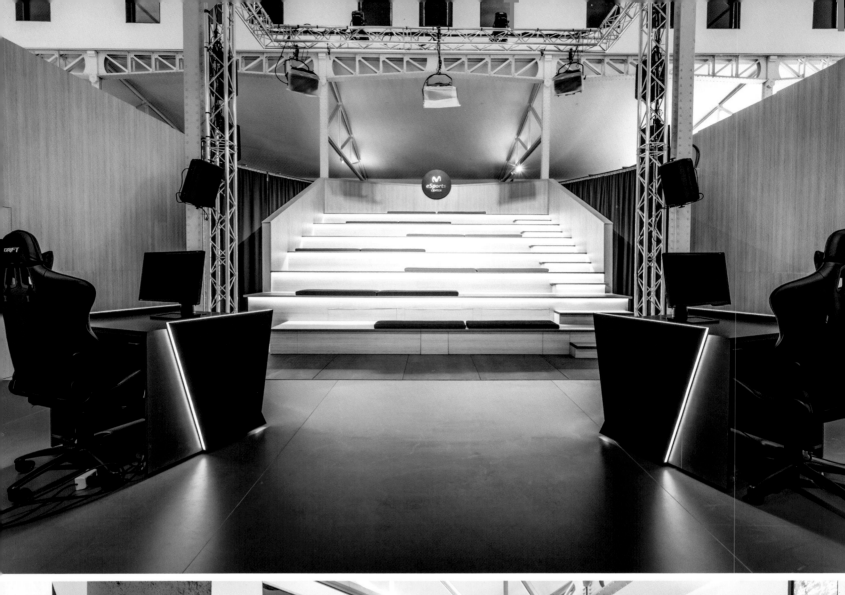

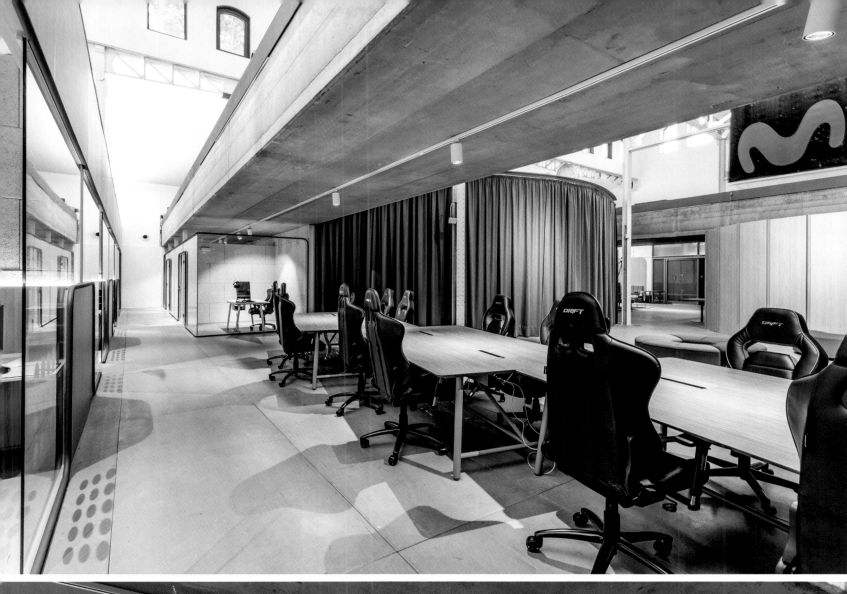

SPACE FEATURES

The project for Movistar Riders is supported by some values that are reflected in the esthetic and functionality of the space that we have created.

The first one is the respect for the building where the project is located. It is a building with a noticeable and symbolic architecture in our city and we wanted to respect it. We have created a series of structures that contrast in materials and co-exist with the building in a harmonic way. They add value to each other, together they emphasize one another. Our project shows respect for our culture and traditions, and it refers to the respect that player will have to others players.

Nowadays, technology is becoming a part of our lives day by day, and so do video games. But there is a time when the technology makes humanity dream with other worlds, unimaginable frontiers and with the idea of getting to farther places than our moon. Video games make us dream bigger than even the NASA engineers.

That time, when the NASA and Russia worked together has allowed us to create an elegant space, but with a retrofuturistic touch. This space allows anyone to feel the warmth of the wood, and also it delights us with organic forms, trapezoidal doors, spaces and seats inspired by those surroundings that were used to create the imagery of the spacial era.

We have created a cozy and functional space so that all team members feel identified and respected. It is a space where the team feeling has to prevail above any other thing. It is a space, with such a strong image, that every single detail breaths the brand and team colors. The new Movistar Riders space is, with no doubt, an elegant place where it is shown that "gamers" are professionals.

Movistar Riders的项目反映了一些价值观，而且这些价值观体现在设计师创建的空间的美观和功能当中。

这是一座在马德里具有明显象征性的建筑，设计师想要尊重它。因而设计师创造了一系列的结构，让它们在材料上形成对比，并以和谐的方式与建筑共存。它们相互增

加价值，并互相强调彼此。也因此，这个项目展现了对马德里的文化和传统的尊重，也饱含着玩家之间的彼此尊重。

如今，科技日益成为我们生活的一部分，电子游戏也是如此。但有一段时间，科技让人类有着各种梦想，如梦想着去到其他世界或从未想象过的地方，或者是去到比月球更遥远的地方。电子游戏使我们的梦想比美国宇航局的工程师都要伟大。

当美国宇航局和俄罗斯共同努力时，我们得以创造了一个优雅的空间，但它带有复古的未来主义的感觉。而眼前这个空间让任何人都能感受到木材的温暖，同时它的有机造型、梯形的门、间隔和座位也让人感到愉悦，这些都被用来创造具有时代意象的空间环境。

设计师创造了一个舒适的功能空间，让所有的团队成员都感觉到被认可和尊重。同时，这是一个团队感受必须凌驾于其他事物之上的空间，也是一个具有强烈形象的空间，每一个细节都要突出品牌和团队的色彩。毫无疑问，新的Movistar Riders空间是一个优雅的地方，它向人们表明"游戏玩家"都是专业人士。

RD CONSTRUCTION OFFICE
RD建设办公室

- Project Name | RD Construction Company
- Design Company | IND Architects
- Designer | IND Architects
- Project Location | Moscow, Russia
- Area | 2,200 m²
- Photographer | Alexey Zarodov

CORPORATE CULTURE 企业文化

Taking City, Future and Innovation as the company's brand foundations, RD Construction is a leading developer of A-grade commercial, industrial and residential projects in Russia. The company is active in Russia, CIS and Europe. Since its foundation, the company has doubled its turnover year on year, with its staff numbers swelling accordingly. Today, RD Construction has over 4,000 professionals on its payroll.

RD建设以城市、未来、创新为品牌基础和口号，是俄罗斯顶级商业、工业和住宅项目的龙头开发商，在俄罗斯、独联体国家和其他欧洲地区都很活跃。自成立以来，该公司的营业额同比增长了一倍，员工人数也相应增加。如今，RD建设已拥有4000多名专业人员。

DESIGN CONCEPT 设计理念

The concept of the office for the building contractor RD Construction is based on its brand foundation and the catchwords. The designer carefully selects the decorative materials and the office furniture, focuses on creating a variety of atmosphere of space, interprets the corporate culture and creates a unique office environment for RD Construction.

建筑承包商RD建设办公室的设计理念基于RD建设的品牌基础和口号。设计师用心选择装饰材料和办公家具，专注营造空间的多种氛围，诠释出公司文化，为RD建设打造出一个独一无二的办公环境。

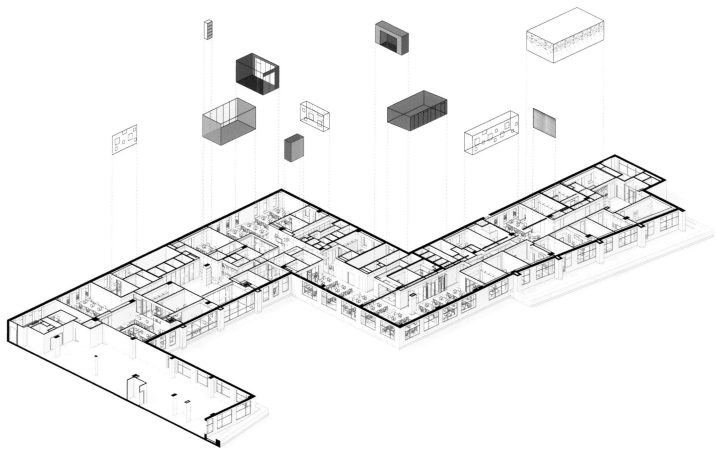

SPACE FEATURES

The catchwords are built into the office's design in the following way: CITY—finishing materials are usually used in facade decor perfectly fit in the office interior. The metal panels, fiber-reinforced concrete and glass allow visitors and company employees to feel like they are in the heart of a megacity inside the office. In addition, these materials eloquently showcase company's operations—among other things, RD Construction has implemented quite a few large-scale urban projects. FUTURE—concise, uncluttered interior, sweeping lines and volumes, and vibrant color accents against an understated color scheme show that here the future is built quite consciously. A column with a media screen in the reception zone makes it possible to show company's projects, including those planned in the future, to clients. The fact that one of the meeting rooms is located near the reception zone accentuates the image of a modern and fast-paced company too. The all-glass volume offers an all-round city view. The volume is separated from the reception zone by a glass wall of a deep turquoise color, which is an accent color in the company's logo. INNOVATION—there are the lamps with a hidden source of light, a polymer cast-in-place floor, the smart glass for meeting rooms that changes its optical properties depending on the privacy requirements and becomes dimmer or fully-transparent, wireless connection between computer hardware and screens. All of these technologies serving as a link between the future and the present, between something ground-breaking and already proven approaches, find their way into the design of the company's office.

Apart from the conventional offices for the management team, the meeting rooms and the open space, there are small meeting points in the office that can be used for recreation or short meetings with colleagues. There is also a meeting room in the VIP zone where more informal meetings with clients and company's management team can be held—the cushioned furniture, diffused lighting, a bar counter, a fireplace, and the heavy curtains create a cozy atmosphere.

RD建设的口号在办公室设计中是通过以下方式实现的：城市——装饰材料，通常用于办公室内部的正面装饰，如金属面板、纤维增强材料。混凝土和玻璃让访客和员工感觉自己在办公室就像处于大城市一样。此外，这些材料在一定程度上展示了公司的运作方式。未来——干净整洁的室内环境，充满活力的色彩亮点与低调的配色方案一起表明这里是非常有意识的建设。在接待区，媒体屏幕的专栏可以向客户展示公司的项目，包括计划中的项目。其中有一间会议室位于接待区附近，突显出一家现代化、快速发展的公司形象。全玻璃的空间提供了一个全方位的城市景观。这个空间与接待区由一堵深绿色的玻璃墙隔开，而这种绿色是公司标志中的突出色彩。创新——暗光源灯、聚合物浇铸的地面，可以根据隐私要求改变光学特性而使其变暗或完全透明的会议室的智能玻璃，计算机硬件和屏幕之间的透明无线连接。所有这些技术作为未来和现在之间的纽带，在一些开创性的和已经被证实的方法之间，找到了融入该公司办公室设计的途径。

除了管理团队、会议室和开放空间的常规办公室外，办公室中还有一些小的会议区，可供娱乐或与同事举行简短的会议。贵宾区里还有一个会议室，铺有坐垫的家具、散光照明、酒吧柜台、壁炉和厚窗帘创造了一种舒适的气氛，在此可以与客户或者和公司的管理团队举行更多的非正式会议。

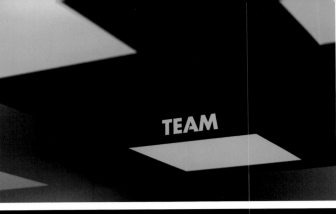

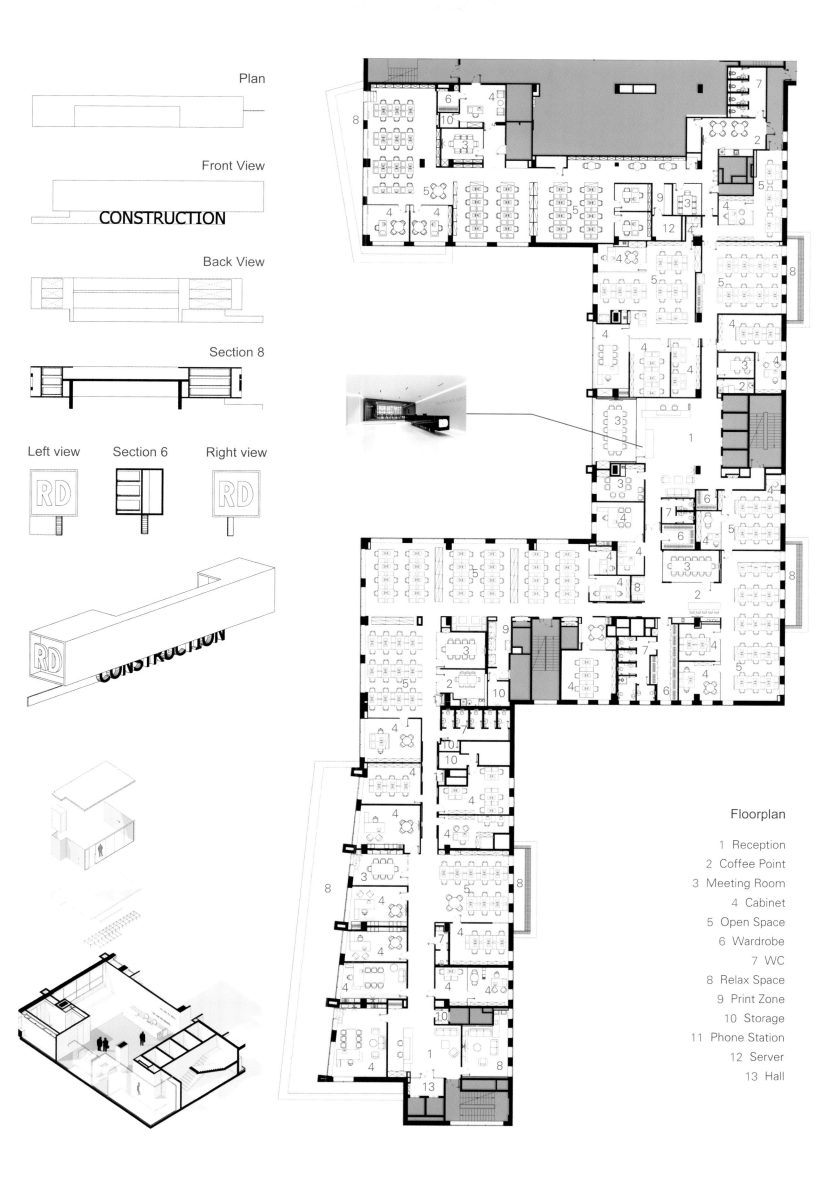

THE SWISS BRANCH OF THE SIGNA HOLDING

SIGNA控股公司瑞士分部

Project Name | The Swiss Branch of the Signa Holding

Project Location | Zurich, Switzerland

Area | 1,000 m²

Design Company | landau+kindelbacher Architekten Innenarchitekten GmbH

Designers | Gerhard Landau, Ludwig Kindelbacher, Christine Sellmeijer

Photographer | Ortwin Klipp

CORPORATE CULTURE

Signa Holding GmbH (officially written as SIGNA) is Austria's largest privately owned real estate company. Signa was founded in 1999 by the Tyrolean entrepreneur Rene Benko. Over the years, the two-man company with an initial focus on classic real estate development has become a pan-European real estate group with more than 150 employees and offices in Vienna, Innsbruck, Munich, Dusseldorf, Zurich and Luxembourg. The principal focus of the Signa Group of Companies is the long-term investment in real estate in prime city centre locations.

SIGNA控股公司于1999年由蒂罗尔州企业家Rene Benko创立，是奥地利最大的私营房地产公司。多年来，这家最初专注于经典房地产开发的小公司已发展成为一家覆盖全欧洲的房地产集团，主要对市中心地区的房地产进行长期投资，拥有150多名员工，并在维也纳、因斯布鲁克、慕尼黑、杜塞尔多夫、苏黎世和卢森堡开设办事处。

DESIGN CONCEPT

This project is located in the "Hochhaus zur Bastei", designed by the famous Zurich architect Werner Stucheli and completed in 1955, and it is considered as the first high-rise building in the city of Zurich. Situated directly on the "Schwanzengraben", the tower block sets an urban development accent and a "gateway" to the city. As an example of high architectural quality, the "Hochhaus zur Bastei" is placed under a preservation order a few years ago. For the Swiss branch of the Signa Holding, the interior design for the new office concept is developed in consideration of the preservation of historical monuments.

该项目位于由苏黎世著名建筑师Werner Stucheli设计并于1955年完工的"Hochhaus zur Bastei"。这座建筑位于"Schwanzengraben"运河边，被认为是苏黎世第一座高层建筑，成为苏黎世城市发展亮点的标志，也是通往该城市的"门户"。作为高层建筑质量的一个例子，几年前，"Hochhaus zur Bastei"被列入建筑保护之中。对于Signa控股公司的瑞士分部来说，新办公室内部设计的概念是在考虑到历史遗迹保护的情况下而开发的。

SPACE FEATURES

Designers use the elegance and generous proportions, together with an unusual and high-quality material mix and accentuated lighting to distinguish the office areas and conference zones in a total of about 1,000m² on four floors. The furniture and fittings are mostly designed and individually installed on site by Landau and Kindelbacher.

With regard to the architectural quality of the Bastei tower and the representative demands of the customer, a special material combination of black marble (polished Nero Marquina), brass, carbon and glass are chosen for the floors, walls and ceilings. As a reminiscence of the architecture of the 1950s, the classical materials are newly interpreted in their use, colorfulness and gloss value. Like a passe-partout, the resulting spatial effect frames the view and allows the low room heights of the existing building to retreat into the background.

In contrast to the black space continua are the high-gloss furniture (highly polished brass) and the golden lighting channels. A speciality is the floor pattern with the brass seams as a visible, high-quality separation. The subject of "seams" is mirrored again and again in the whole design—from the integration of direct and indirect lighting in the cooling ceiling to the special furniture.

该办公室总面积约1000m²，共占据四层楼。设计师用优雅大方的比例，搭配别致的、高质量的材料和突出的照明灯具，从而区分办公区域和会议区域。家具和配件大多是由Landau和Kindelbacher设计，并在现场独立安装完成的。

针对Bastei的建筑质量和客户的需求，地板、墙壁和天花板由黑色大理石(抛光的黑白根大理石)、黄铜、碳和玻璃等特殊材料组合而成。作为对20世纪50年代建筑的一种追忆，古典材料在使用、色彩和光泽度等方面都得到了新的诠释，就像一个画框所产生的空间效应可以框住景色，这里让现有建筑的较低的房间成为背景。

与黑色空间形成对比的是高光泽的家具(高度抛光的黄铜)和金色的照明通道。地板使用黄铜接缝作为一种可见的、高质量的分隔线，成为整个空间的一个特色。从冷却天花板到特殊家具的直接和间接照明的融合，整个设计一直在反映"接缝"的主题。

KOCHANSKI ZIEBA & PARTNERS LAW FIRM

Kochanski Zieba & Partners律师事务所

Project Name | Kancelaria Kochanski Zieba & Partners

Project Location | Cracow, Poland

Area | 1,400 m²

Design Company | Iliard Architecture & WProject Management

Designers | Wojciech Witek, Lukasz Koziana, Katarzyna Smagala, Katarzyna Miodek, Krzysztof Lesniewski

Photographer | Bartlomiej Senkowski

CORPORATE CULTURE 企业文化

Kochanski Zieba & Partners is one of the most renowned in the country, and currently consists of more than 100 lawyers, legal counselors, tax counselors and patent attorneys co-operating through offices in Warsaw and Cracow. We are a full-scope, business orientated law firm dedicated to providing clients with innovative and efficient legal solutions of the highest professional standards in various practice areas.

Kochanski Zieba & Partners是波兰著名的律师事务所，目前由100多名律师、法律顾问、税务顾问和专利代理人通过在华沙和克拉科夫的办事处合作组成。这是一家全方位的，以业务为导向的律师事务所，致力于为客户提供创新、高效的法律解决方案，在各个领域实践最高的行业标准。

DESIGN CONCEPT 设计理念

The aim of the architects, right from the beginning of conceptual works concerning this project, is to bring together elegance and timelessness while at the same time underlining an expressive and clean-cut character that is crucial for achieving success in the law business. A distinctive, ultra-modern reception desk in pale white is contrasted with black construction profiles and glass walls. The stern atmosphere of the office is toned with the warmth of wooden veneer walls. Those are the main elements that give this space a certain unmistakable touch and make in etch in your memory.

建筑师们从这个项目的概念工程开始，将优雅和永恒相结合，同时强调空间表现力和冷静理性的性格，这是取得成功的法律业务的关键。一个造型独特的、超前的白色接待处与建筑黑色轮廓和玻璃墙形成鲜明对比。办公室的严峻气氛被温暖的木质贴面墙调和。这些为这个空间提供一种明显的触摸感，是让你有深刻的记忆的主要元素。

SPACE FEATURES

A seemingly levitating reception desk is manufactured using the micro topping technology. The outcome is a solid, uniform piece. At the same time, thanks to raising it up from the ground in a sort of illusory manner, the light and subtle effects are reached. Aside from that, architects design the mobile walls and ample openings in the areas reserved for the meeting rooms. This move enables the quick transformation of those spaces together with the foyer into a substantial event room with a spectacular view onto three directions of the city.

Achieving the formal spaces made-to-measure of the demands of a lawyer's team is a challenge to the architectural crew. Simultaneously, they are determined to highlight individual schemes dedicated to this particular project, its distinguishing elements. The outcome of these efforts seems to exceed the expectations.

一个看似悬浮的接待台是使用微型技术制造而成的，是一个坚实的、完整的作品。设计师以一种虚幻的方式将它从地面抬起，从而得到一种轻便而微妙的效果。除此之外，建筑师设计了可移动的墙壁，并为会议室预留了宽敞的开放空间。墙壁这一设计能够将这些空间与门厅快速转变为一个实质性的活动室，并可欣赏到城市三个方向的壮观景色。

要达到律师团队对这个正式空间的设计要求，对建筑团队来说是一个挑战。因此，他们决定专门为这一特定项目做针对性的有特色的设计方案。而最终，这些努力收获到了超出预期的结果。

PARADOX INTERACTIVE HEADQUARTERS

Paradox Interactive 总部

Project Name | Paradox Interactive

Design Company | Adolfsson & Partners

Designer | Adolfsson & Partners

Project Location | Stockholm, Sweden

Area | 3,312 m²

企业文化
CORPORATE CULTURE

Paradox Interactive was formerly a division of Paradox Entertainment, rights holders of properties such as the Robert E. Howard character Conan. Now, it is a Swedish video game publisher based in Stockholm. The company is best known for releasing historical strategy video games. Paradox Interactive publishes its games, both developed by their division, Paradox Development Studio, and those of other developers.

Paradox Interactive以前是Paradox娱乐公司旗下的一个部门，拥有诸如Robert E. Howard等作家的版权。现在它是瑞典的一家视频游戏发行商，总部设在斯德哥尔摩，以发行历史策略类的电子游戏而闻名。同时，Paradox Interactive也发行自己子公司开发的游戏及其他公司开发的游戏。

设计理念
DESIGN CONCEPT

The people working for this company are dedicated gamers themselves, and could be referred to as nerds. We decided to embrace that, putting references, characters and scenery from Paradox's different games merged into the interior design in a subtle way. Graphic illustration, hexagon shapes and striking photo covered walls as backdrops are put together with nerdy codes and stunning space visuals. The result is an office for multiple players and a lot of winners.

在这家公司工作的人都是可以被称为"网虫"即专注于游戏的人。设计师决定接受这一特点，将Paradox不同游戏的标志、人物和风景以一种微妙的方式融入室内设计中。墙壁上的图形插图、六边形形状和醒目的照片作为背景，与"网虫"代码和令人惊叹的太空视觉结合在一起，最终为许多玩家和赢家提供了一个办公室。

扫码查看电子版

HUMANE OFFICE 人文办公

SPACE FEATURES

The office is 3,312 square meters spread over three floors, containing 274 workstations, 19 meeting rooms, a sound studio, a game testing studio and a striking training room in the form of a church which we named "The Paradox Chapel".

The color scheme is consistently in the black, gray and white accented with muted colors in red, yellow, blue and green, directly downloaded from the Paradox game graphics.

In the entrance, you are met by a central reception desk in the form of an external hexagon. The hexagon form will return later in the pendant luminaire above, where 150 pcs hexagonal modules is shaped into an organic honeycomb structure.

On the wall behind the front desk hailed Paradox games portfolio with a large colorful artwork in the form of a collage and on the meeting rooms glass panels, you see the black line art illustrations. The office landscape is separated by the meeting room whiteboard walls which creates the smaller working groups. Each whiteboard is framed with actual code strings from Paradox games.

Pillars and walls adorned with flat painted quotations, both well-known but also internal and in foreign languages. Employees' lockers have been treated and given a patina to resemble those in military facilities. Each lounge area which is interspersed with the office landscape has its special feel and clear theme with inspiration from the games "Hearts of Iron", "Magicka" and "Stellaris". The dining room is a place to meet and gather for big events. The furnishing is varied with mostly picnic tables, cozy booths and a long bar counter for the Friday beer and a stage for presentations.

办公室占地3312㎡，共三层楼，包括274个工作点、19个会议室、一个音响工作室、一个游戏测试室和一个教堂式的训练室，设计师称之为"The Paradox Chapel"。

配色方案延续了黑色、灰色和白色，同时搭配运用Paradox游戏中所使用的颜色，如红色、黄色、蓝色和绿色，令其成为整个设计的亮点。

在入口处，你可以看到一个六边形的中心接待桌。上方的吊灯延续了六边形的形状，用150个六边形小模块组合成一个有机蜂窝结构。

在前台后面的墙上，拼贴的大的彩色艺术品呈现着Paradox的游戏组合，以及在会议室的玻璃面板上，你可以看到黑色线条艺术插图。办公室景观被会议室的白板墙隔开，从而形成了小的工作组。每个白板上都有来自Paradox游戏的实际代码字符串。

在柱子和墙壁上，设计师用瑞典语和外语表示的著名格言作为装饰。员工的个人储物柜是类似于军事设施的样式。每个休息区都有来自游戏"钢铁雄心""魔法对抗"和"群星"的独特的感觉和清晰的主题。餐厅是一个聚会的地方。餐厅里的家具种类繁多，主要是野餐桌、舒适的隔间以及为每周五的啤酒会所开设的吧台和演讲舞台。

INCANTO MOSCOW OFFICE
Incanto莫斯科办公室

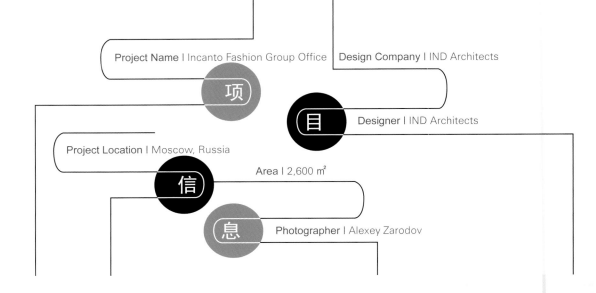

Project Name | Incanto Fashion Group Office
Design Company | IND Architects
Designer | IND Architects
Project Location | Moscow, Russia
Area | 2,600 ㎡
Photographer | Alexey Zarodov

企业文化 / CORPORATE CULTURE

Incanto is an emerging fashion brand of lingerie in Italy and one of the leading brands of lingerie in the world. It also has two brands, Innamore and Malemi, and its business includes the production and sale of home clothes, underwear and hosiery. The design of its products is full of interesting elements, close to the latest fashion trends and strives to blend the traditional features of Italy with the current trends.

Incanto是意大利一个新兴的时尚内衣品牌，也是全球领先的内衣品牌之一，旗下还有Innamore和Malemi两个品牌，业务包括家居服、内衣和针织品的生产和销售。Incanto的产品设计充满奇趣，并紧跟时尚潮流，力求把意大利的传统特色与当前流行趋势融合在一起。

设计理念 / DESIGN CONCEPT

A Moscow office that occupies three stories of a business center is home to three companies of one holding. This is a contemporary-styled interior inspired by nature. Austere and uncluttered lines are complemented with the ingenious patterns that add individuality to each office.

Each office of the holding was realized in an individual stylistics and color scheme matching its business line. The interior of the first company is made in light shades that smoothly match warm deep colors. The second office is full of austere geometric lines, intense dark colors and bright shades. The third office, too, is made in dark colors. The fashion industry is represented by the highlighted bright posters on the walls, a glassy metal perforated ceiling and a reception desk that looks like a piece of a crystal.

这套位于莫斯科的办公室占据了一个商业中心的三层楼，它是三合一的控股公司。办公室内现代风格的设计灵感来源于大自然，简朴整洁的线条加上巧妙的图案赋予不同的办公室独特的个性。

三家公司，每一家的办公室都是按照其个人风格和配色方案来设计的，使办公室的风格与该公司的业务相符。第一家公司的室内设计运用柔和色调与温暖的深色调相呼应；第二家公司的办公室充满了朴实的几何线条，强烈的深色和明亮的色调；第三家公司的办公室同样也是深色的，通过墙上明亮的海报、玻璃金属穿孔的天花板和一张看上去像水晶的接待台来体现时装行业的时尚气质。

空间特色
SPACE FEATURES

Natural stone, metal, glass, wood and ceramics are used in the finishing of the office. Natural green walls of live plants with an automatic watering system add brightness and harmony. The pattern of air is used on glasses in the open space of the "ground" part of the office—a light-colored pattern adds lightness and aeriality to the space. The wall behind the reception desk is made of top-lit transparent sheet glass. Such solution allows visitors to enjoy a beautiful view from the window and does not impede the penetration of light into the hall space.

There is a small common room in the office, where one may switch off from work, read books and magazines. Cushioned furniture, a cozy stand lamp, a green shag carpet that looks like grass and a gray timber wall with an unusual pattern of bright green lines representing life, help to lay back nicely in a matter of minutes.

办公室的装修使用了天然石头、金属、玻璃、木材和陶瓷。植物形成的天然绿色墙壁和自动浇水系统可增加空间亮度与和谐度。空气纹理被运用在开放空间的"地基"部分的玻璃上，其浅色图案增加了空间的亮度和空虚感。工作间内部整洁且具有现代感，带有少量的色彩。接待处后面的墙壁是用顶部亮着灯的透明玻璃做的，这样的设计可以让访客从窗口欣赏到美丽的景色，也不妨碍光线进入大厅。

办公室里有一间小休息室，人们可以在这里休息、看书、翻阅杂志。带有靠垫的座椅、一盏舒适的台灯、一张看起来像草地的绿色地毯，还有一堵灰色的木墙，上面画着代表生机的特别明亮的绿色线条，这些元素都有助于员工在极短的时间内便可很好地进入休息状态。

Floor Explication

1　Reception
2　Relax Space
3　Meeting Room
4　Open Space
5　Cabinet
6　Storage
7　WC
8　Wardrobe
9　Coffee Point
10　Server
11　Showroom

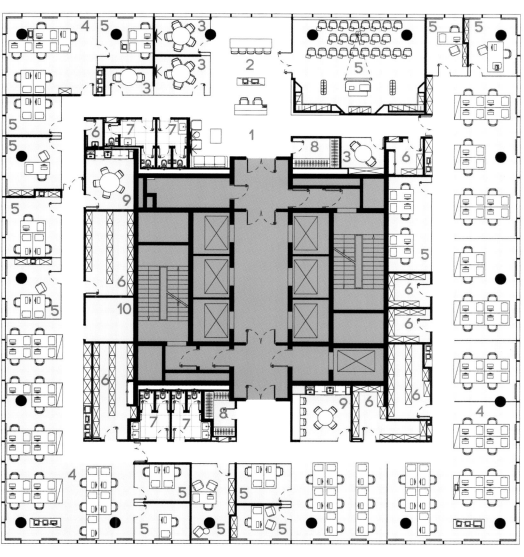

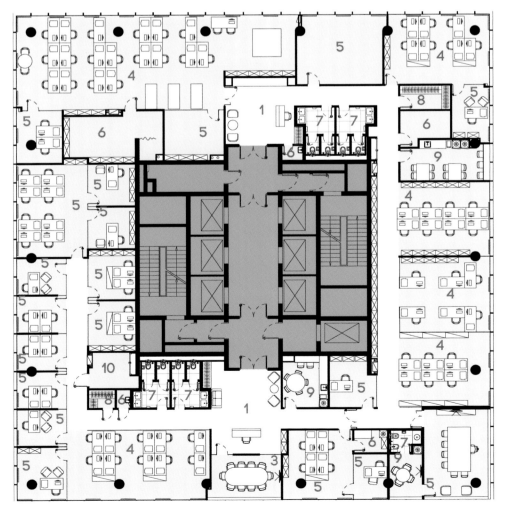

Floor Explication

1 Reception
2 Relax Space
3 Meeting Room
4 Open Space
5 Cabinet
6 Storage
7 WC
8 Wardrobe
9 Coffee Point
10 Server
11 Showroom

Floor Explication

1 Reception
2 Relax Space
3 Meeting Room
4 Open Space
5 Cabinet
6 Storage
7 WC
8 Wardrobe
9 Coffee Point
10 Server
11 Showroom

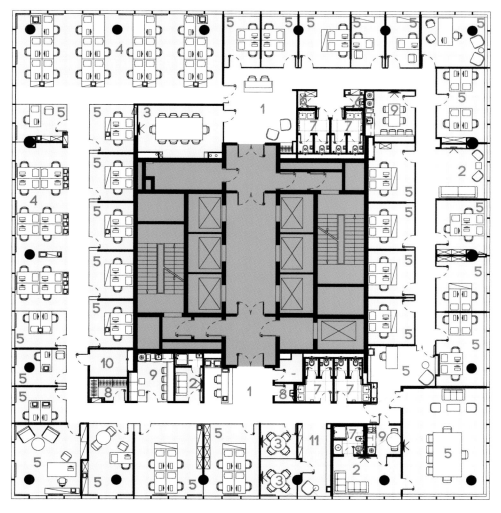

ILGE BOOKSTORE OFFICE
ILGE的书店办公室

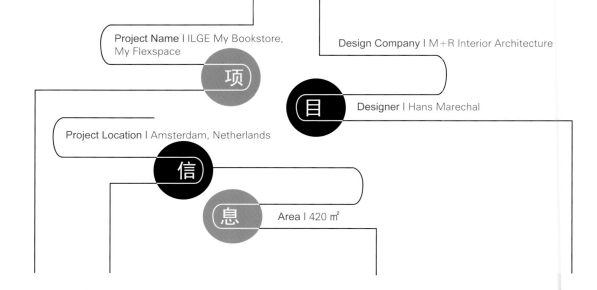

Project Name | ILGE My Bookstore, My Flexspace
Design Company | M+R Interior Architecture
Designer | Hans Marechal
Project Location | Amsterdam, Netherlands
Area | 420 ㎡

CORPORATE CULTURE

ILGE is a leading player in subscription management, online content management and collection advice on the Dutch, Belgian and German markets. In the thirty years of its existence, ILGE has found the ideal mix between traditional, customer-friendly service and surprisingly innovative solutions. From July 2017, ILGE moved her desk to Villa Mokum in Amsterdam, a building with studios and a plinth with retail spaces on the Spaklerweg. A new area development for living and working. In addition to the design for the innovative work environment for ILGE, there is also a new bookstore annex flexspace concept housed in the retail space of a total of 420 ㎡.

ILGE在荷兰、比利时和德国市场的订阅管理、收集咨询方面处于领先地位。在其存在的三十年中，ILGE找到了在传统的友好服务和新颖的创意方案之间的理想组合。2017年7月，ILGE搬进阿姆斯特丹的Mokum别墅，这是一幢位于斯帕克里维斯的带有工作室和零售空间的建筑，为生活和工作开辟了一个新的领域。除了为ILGE设计的创新工作环境之外，这里还有一个附带灵活空间概念的新书店，位于420㎡的零售空间中。

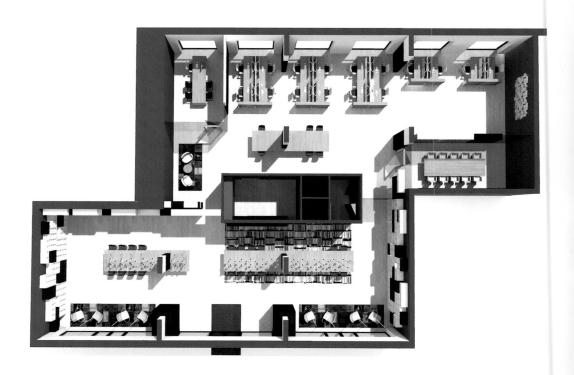

DESIGN CONCEPT

The book market has been moving for years. The share of e-books continues to grow, from the total sales, 6% is being sold as an e-book. The proportion of e-books in online sales is considerably higher: at 28% and these percentages increase. There are also several larger (online) providers who are now in the book sales market. This development will continue in the coming years. Additionally, the increasing branch evolution also plays a major role in the declining turnover of the traditional bookshops. We see that supermarkets, gas stations and department stores also sell books. With this knowledge of the book market, we have developed a completely new interior concept with our client, with the work-title "the last bookstore".

图书市场多年来一直在变化。电子书的份额持续增长，从总销量来看，6%的书籍是作为电子书销售的。电子书在网上销售中所占的比例要高得多，达到28%，而且这些百分比还在增长，还有几家更大的供应商（线上）驰骋在图书销售市场上，这一发展趋势将在今后几年持续下去。此外，传统书店营业额的下降，日益增加的分支机构的演变在这里发挥了重要作用，我们可以看到大卖场、加油站和百货公司都在卖书。通过对图书市场的了解，我们与客户一起开发了全新的室内概念，并以"最后一家书店"为命名。

SPACE FEATURES

Book stores will have to respond more than ever to the rapidly changing needs in the market. The concept for this bookstore has become much more a meeting place, a flexspace where one can work, meet and acquire knowledge between the "empty" bookshelves in an attractive space. Because the book world of adventure, knowledge and science is not black and white, we have created a photo print of a huge colorful bookcase for the middle part: it covers both the floor, wall and ceiling and is the only color accent in space. The concrete structure and installations are "pure" and visible in space. A flexible light concept with tracks and the adjustable led spots make the atmosphere. The bookstore and office are separated by a facility unit where the kitchen, toilets and engineering rooms are housed. A special website has been developed, one can share working and meeting in the bookstore with others, have a healthy lunch or book a meeting room.

Because bookstores need to be part of the online buying and orientation behavior of the consumer, iPads are included in the tables that allow the visitor access to an inexhaustible database with both international and local magazines, newspapers and books. The publications can be viewed here and ordered as an extra service: the E-bookstore. The challenge is whether the bookcases remain empty or still used to show a collection, this is one of the possibilities.

书店将比以往任何时候都更需要对市场上瞬息万变的需求做出对策。这是一间书店，更是一个会议场所、一个互相吸引的空间，人们能够在"空荡荡"的书架之间工作、见面和获取信息。位于房间中央的大桌子是几个工作区和咨询区，一些双门跑车一样的座位区域集成到橱柜的墙壁。因为关于冒险、文学和科学的书的世界不是黑白的，所以设计师在中间部分创造了一个巨大的彩色书柜照片：它涵盖了地板、墙壁和天花板，是整个空间中唯一的亮点。具体的结构和装置都非常"干净"，在空间上是可见的。此外，灵活的、可调节的灯光可以渲染气氛。书店和办公室由厨房、洗手间和工程室组成的设备单元隔开。我们开发了一个专用的网站，在书店，你可以和别人分享工作和开会，享用一份健康的午餐或者预订一间会议室。

因为书店需要成为消费者在线购买和咨询的一部分，所以iPad被安装在桌子上，允许顾客访问这取之不尽的数据库，里面包含有国内外的杂志、报纸和书籍等。这些出版物都可以在这里浏览，并作为额外服务在电子书店订购。而目前，书柜是否要保持为空的，还是照旧用来展示书籍，这是一个值得考虑的问题。

LOW-CARBON OFFICE 低碳办公

- 如果您拥有一个创造梦想的空间
- 可以这样装扮它
- 也可以这样装扮它
- 还可以这样装扮它
- 在这里，你总能找到属于自己梦想的办公空间

THE OXYGEN BAR: BEIJING ZHONGJUN WORLD CITY FUNWORK
丛林氧吧：北京中骏世界城FUNWORK

项目信息

- 项目名称 | 北京中骏世界城FUNWORK
- 设计公司 | VHD维度华伍德设计机构
- 设计师 | 廖建锋、谢圣海、钟江晓、陈文良、陈祖川
- 项目地点 | 中国北京
- 项目面积 | 5,300 ㎡
- 项目材料 | 橡木、雅士灰石材、美岩板、素水泥、红砖等
- 摄影师 | 上海金选民摄影有限公司

企业文化 / CORPORATE CULTURE

FUNWORK is a member of Shanghai Leyao Venture Capital Management Co., Ltd. The founding team is a group of overseas returnees born in 1980s, who are the ideal but down-to-earth young people and have experiences in the marketing and the merchants operation of more than 40 projects of the famous national real estate enterprises for many years. FUNWORK is dedicated to upsetting the traditional, boring and stressful way of working, creating a free and new concept of office space to help entrepreneurs realize their dreams. FUNWORK also provides a full range of business training, policy applications, business registration, finance and other office services, dedicated to create an entrepreneurial ecological community system.

FUNWORK隶属于上海乐耀创业投资管理有限公司，创始团队是一群80后的"海归"，有理想抱负且脚踏实地，拥有全国知名房地产企业四十余个项目营销、招商运营的多年经验。FUNWORK致力于颠覆以往传统、无趣、压力大的工作方式，打造一个自由的且是全新概念的办公空间，让工作遇见自然，帮助创业者实现自己的梦想。另外，FUNWORK还提供创业培训、政策申请、工商注册、财务等全方位办公服务，试图打造一个创业生态社群体系。

设计理念 / DESIGN CONCEPT

Beijing Zhongjun World City FUNWORK is the first project that makes FUNWORK into Beijing's market. Huang Tao, the founder, is determined to overturn the closed, boring and depressed working mode and creates the concept of the biggest business, office ecological community in China. Under the high-stress condition of the entrepreneurial environment, FUNWORK strives to present a happy, shared, open and different ideas of the joint office atmosphere and make staff can breathe and decompress here. In consideration of customers and entrepreneurs, designers create this natural ecological oxygen bar in Beijing, the jungle of iron and steel.

北京中骏世界城FUNWORK是FUNWORK进军北京的首个项目，秉承创始人黄涛立志要颠覆封闭、压抑的工作方式以及打造全国最大的创业、办公生态社群的概念，力求在现在高压力的创业生存环境之下，表现一个欢乐、共享、开放、不同思想碰撞的联合办公氛围，并能让工作者在此呼吸、减压。在考虑客户和创业者的前提下，设计师在北京的钢铁丛林中，打造出这个自然生态的办公氧吧。

263

SPACE FEATURES

The use of a large number of original ecological planks and green planting walls makes the whole space close to the original ecology, creating an easy and casual office environment. The original white bearing column without any decorations also is decorated by the wooden finishing and evolves into a towering tree. The branches stretch to the ceiling, on which the tree houses are hanging, adding the vitality of nature to the monotonous ceiling.

Slide, swing chair, customized personality office furniture, books bar, coffee area, water bar, massage room, pool table and other entertainment function are integrated into the office space. Thus, the boundaries between entertainment and work are no longer distinct, but the mutual fusion and inclusion of each other. It breaks the traditional rules of the past, and is represented in the form of a more rapid rendering.

This space is full of the natural flavor, and the childish interest is also booming in this natural atmosphere. Shuttling through space is like walking in a lush but uncluttered forest, where has the slide in the form of sculpture, hobbyhorse recalling the Trojan Horse, the rocking chair in the meeting room and the mathematical symbol "∞" in the brainstorming area and others. All of these elements can remind people's initial heart, make people keep childish and visionary imagination and mean the opportunity and development space of the career.

大量原生态木板、绿植墙的采用使整个空间接近原生态，营造出轻松、休闲的办公环境。过去毫无装饰的白色承重柱，也被木制包裹，演化成了一棵参天大树。树枝向天花板伸展开来，一个个树屋垂挂其中，给单调的天花板中加入了自然的生气。

滑滑梯、秋千椅、定制的个性办公家具、书吧、咖啡吧、水吧、按摩室、台球桌等休闲功能大量融入办公空间当中，休闲和工作的界限不再泾渭分明，而是相互融合、彼此包容，打破以往中规中矩的束缚，以一种更具跳跃性的形式呈现。

空间中的自然气息到处弥漫，无处不在的童趣也在大自然般的氛围中蓬勃迸发。穿梭在空间里，宛如走在枝繁叶茂却不杂乱的森林中，里面以雕塑的形式打造的滑滑梯、引人联想到特洛伊战争木马计的木马、会议区的摇椅、头脑风暴区的数学符号"∞"造型等，都让人想起自己的初心，让人保持童心、天马行空的想象力，也寓意事业的机遇和发展空间。

KING KUNGSGATAN 36 OFFICE

King 国王街36号办公室

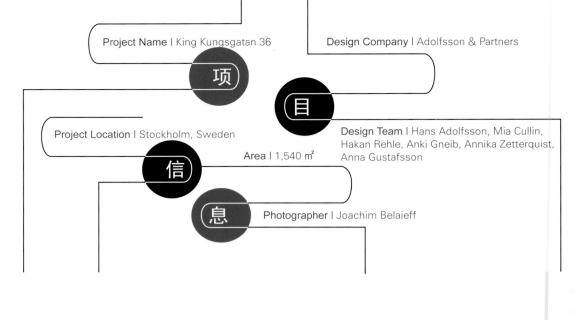

- **Project Name** | King Kungsgatan 36
- **Design Company** | Adolfsson & Partners
- **Project Location** | Stockholm, Sweden
- **Area** | 1,540 m²
- **Design Team** | Hans Adolfsson, Mia Cullin, Hakan Rehle, Anki Gneib, Annika Zetterquist, Anna Gustafsson
- **Photographer** | Joachim Belaieff

CORPORATE CULTURE 企业文化

King is one of the world's leading mobile interactive entertainment companies, and we have gamers all over the world. We have developed more than 200 interesting works, and these games have been launched in many countries and regions of the world. King has created several hundred games, and even though they are different, they all take place in various fairytale landscapes.

King是世界领先的手游公司之一，游戏玩家遍布世界各地。他们已经开发了超过200种充满乐趣的作品，这些游戏已在世界多个国家和地区推出。King创造了几百款游戏，尽管它们各不相同，但它们都发生在各种童话场景中。

DESIGN CONCEPT 设计理念

For this new project, we go back to the roots with the inspiration sprung from the Swedish nature. Every project has its challenges. But with the right people, innovative solutions and technical excellence, we manage to work alongside the disadvantages and make them a part of the concept. King's games inspired us when we divided the kingdom into ten different areas: Deep Sea, Countryside, Treasure Island, Green Hills, Magic Forest, Sandy Dunes, Mountain Tops, Wild Jungle, Kingtown (Upper and Lower) and Pavilion Park.

关于这个新项目，设计团队将设计回归了源于瑞典自然的灵感根源。每个项目都有其挑战性，但是有合适的人才、创新的解决方案和卓越的技术，设计团队便能设法克服困难，并将这些优势特点糅合成了设计理念的一部分。King的游戏王国激发了设计团队的灵感——将这个办公室王国划分成十个不同的区域：深海、农村、金银岛、青山、魔幻森林、沙丘、高山、野生丛林、国王镇和亭阁公园。

 空间特色
SPACE FEATURES

To make the most of the natural light in this studio, we make the glass-roofed as well as a local hero in The Forest, the centerpiece of this office. Rolling hills, a wall covered with lichen, hanging basket-chairs are some of the features completing the strong vision for this studio. Combine that with unique solutions, such as the interactive floors making streams turn into ice as the season changes.

Like a plant needs a good place to grow strong, so does a good idea. With that in mind the project grows into something that would focus on how the creative mind evolves with the recreation as well as the inspiration, which is essential to accomplish the great work and succeed in the online gaming industry. As a result of this project, Adolfsson & Partners gives King an office space where ideas grow big and better.

  为了充分利用这个工作室的自然光线，设计师设计了一个玻璃屋顶，而办公室的中心则设计了一位当地的英雄在"森林里"。起伏的山丘、覆盖着地衣的墙壁、悬挂的摇椅是完成这个工作室强烈愿景的一些特色，结合独特的解决方案，如互动地板可以使溪流的水会因冬季的到来而结冰。

  就像植物生长需要一个好的环境，一个好的想法同样需要好的工作环境来培育。考虑到这一点，这个项目被发展成为一种专注于创造性思维是如何随着娱乐和灵感的发展而发展的场所，而灵感对于伟大工作的完成和网络游戏产业的成功发展是必不可少的。因此，设计团队为King提供了一个可以使创意变得更好、更伟大的办公空间。

KING SVEAVAGEN 44 OFFICE

King Sveavagen44号办公室

282/283 LOW-CARBON OFFICE 低碳办公

Project Name	King Sveavagen 44
Design Company	Adolfsson & Partners
Designer	Adolfsson & Partners
Project Location	Stockholm, Sweden
Area	6,578 m²

CORPORATE CULTURE 企业文化

King is one of the world's leading mobile interactive entertainment companies, and we have gamers all over the world. We have developed more than 200 interesting works, and these games have been launched in many countries and regions of the world. King has created several hundred games, and even though they are different, they all take place in various fairytale landscapes.

King是世界领先的手游公司之一，他们的游戏玩家遍布世界各地。他们已经开发了超过200个充满乐趣的作品，这些游戏已在世界多个国家和地区推出。King创造了几百款游戏，尽管它们各不相同，但它们都发生在各种童话场景中。

DESIGN CONCEPT 设计理念

Detailed preliminary studies of the employees' wishes resulted in more than just the idea that the office should "live and breathe King". People wanted an open, easily navigated office landscape with the clear dividers for the various project teams. There was also a desire for many lounges, meeting rooms and a variety of creative spaces.

Using this information as a foundation and with an ambition to bring King's world of games and characters to life, we created a colourful and energy-filled office featuring both humor and intelligent solutions.

设计团队对员工的愿望进行了详细的研究，得出的结果是办公室不只是"生存和呼吸"的地方。人们想要一个开放的、功能区明确的办公室，并希望能为不同的项目团队明确划分空间区域。他们还渴望有许多休息室、会议室和各种创意空间。

以这些信息作为基础，我们创建了一个既幽默又聪明，色彩丰富且充满活力的办公室，就是为了让King的游戏世界和人物生活充满活力。

287

SPACE FEATURES

The task of this project is to create an office where something new can be discovered around every corner. A creative office landscape that communicates King's soul, a place that with "fun" and "magic" as its watchwords can be called a kingdom.

Each landscape has been carefully and consistently coded and staged to ensure recognition and easy navigation by colors. Graphics from the world of games are constantly presented and playfully applied as wallpaper, vinyl on glass or three-dimensional objects.

This is an office that reflects King—what they create and what they believe in. It is out of the ordinary, just like King.

本项目的任务是让人可以在任意一个角落都能发现一些新东西。这里有创意十足的办公室景观，能够传达出King的灵魂，它是一个具有"乐趣"和"魔力"的地方，恰如它的口号——王国。

办公室的每一处景观都是使用严谨且一致的颜色进行编码和分级，易于识别和导航。游戏世界的图形不断呈现，富有玩味儿地应用在壁纸、玻璃上的乙烯基或三维物体上。

这是一个反映King的创造和信仰的办公室，就像国王一样与众不同。

THE LEGO SPIRIT OF THE CONSTRUCTION COMPANY

建筑公司的乐高精神

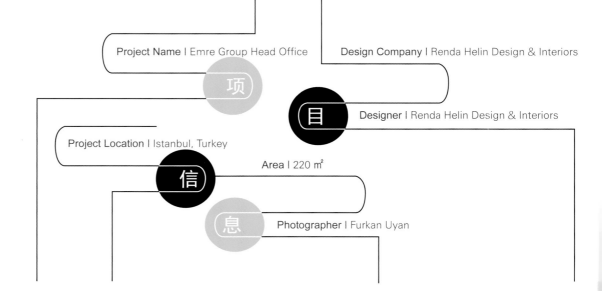

Project Name | Emre Group Head Office
Design Company | Renda Helin Design & Interiors
Designer | Renda Helin Design & Interiors
Project Location | Istanbul, Turkey
Area | 220 m²
Photographer | Furkan Uyan

CORPORATE CULTURE

Emre Group is established with its visionary approach to the construction, tourism and automotive sectors. The foresight and the innovative perspective of the corporation come to life in the interior design. EMRE GROUP, which is located at the largest metropolitan city with its historical and cultural texture, has signed the big urban transformation projects and started to provide services in the fields of Real Estate Development, Construction and Architecture with its group companies since 2016. Our mission is to build modern and contemporary buildings that combine modern interior and exterior architecture with comfort.

Emre集团是基于其在建筑、旅游和汽车行业的远见成立的，公司的远见卓识和创新视角在室内设计中焕发出生机。Emre集团位于具有历史和文化底蕴的大都会城市，自2016年起与大型集团公司签署了大型城市改造项目，并开始在房地产开发、建筑和建筑领域提供服务。公司的使命是建造现代化的建筑，把现代的室内和外部建筑与舒适感结合起来。

DESIGN CONCEPT

The firm has been serving extensively in the construction industry, which constituted the main concept for the interior design. The moment you step into the building, the touches that make you feel like you are on a construction site are remarkable. The industrial design approach is adopted in order to represent the construction sector, and is reflected especially in the reception and accountancy department, the director room, the meeting room and the entrance floor where the common WC is located. It is also further enhanced by the yellow and black colors which are also another reference to the construction sector. The material preferences are determined by the raw and natural appearance of the building materials. The indispensable elements of the construction atmosphere are revealed by the use of materials such as concrete, epoxy, brick, iron and mesh metal.

LOW-CARBON OFFICE 低碳办公

303

该公司在建筑行业的广泛发展为设计师提供了设计思路。当你踏进建筑的那一刻，就好似进入到一个非同寻常的建筑工地。为了体现建筑行业，设计师采用了工业设计的方法来体现建筑行业，尤其突出在接待区、财务室、主任室、会议室和一楼公共卫生间的入口等区域。在建筑工地上经常出现的黄黑色警示色彩也在设计中得到强化。材料的选择取决于建筑材料的原材料和自然外观，设计团队选用在建筑行业中不可或缺的常用材料，如混凝土、环氧树脂、砖块、铁和金属网等。

空间特色
SPACE FEATURES

One of the most prominent points of the project is undoubtedly hidden in the details. *The Lego Movie* inspired the design of the meeting room. The movie's main motto, "Grab a hat let's build something" has become the basic vision and slogan of the company. A part of the meeting table was made of the original legos, while the other part is made with a lighting element designed specially in the form of a crane. The hammers placed in the wall of the meeting room represent the constructions of Emre Group. At the same time, the hammer symbolizes the force, labor, power of work and the whole state of existence of a building. Each hammer has the district name, parcel and island number referring to the location of the project. Graphic works come forward in design, and a slogan expressing the worldview, philosophy and vision of the company can be seen on the gross concrete wall of the staircase.

On the lower floor of the building, there is a hobby room which will be integrated with the technical staff room, kitchen, staff rest rooms, executive office and the inner courtyard. Each space is designed to represent its concept. The company owner's room is designed in black and white color scheme, together with raw and natural materials as in the other parts. The executive table is designed and manufactured to serve as meeting table as well, and to meet all expectations of the user.

The indoor courtyard where an ornamental pool and a tree are located. Water represents abundance, whereas tree represents the company sensitivity for the environmental worries. The hobby room, integrated with the inner courtyard, represents the background of the company owner as a black belt owner in boxing and karate. It has been designed to provide exercise opportunity in the office.

这个项目最突出的特点都隐藏在各种细节中。《乐高大电影》的创意思维启发了会议室的设计。电影的主题句"拿起帽子，让我们做点什么"已经成为公司的基本愿景和口号。会议桌的一角由彩色乐高积木拼贴而成，另一部分的照明灯具来自大吊车的构造元素。会议室墙壁上的椰头代表着公司在建筑领域的印迹，与此同时，椰头还象征着建筑的力量、劳动力、工作动力和整个行业的生存状态。每一把椰头分别根据项目所在地加以命名和编号。从本案的图形设计中，楼梯间混凝土墙面上的文字标语表达出公司的价值观、哲学观和愿景。

一楼有一个休闲活动区，这个活动区是将技术人员室、厨房、工作人员休息室、行政办公室和内部庭院相结合而成的。办公室中每个空间都有自己的理念：公司老板的办公室采用黑白配色方案，搭配其他原材料和天然材料；设计团队特别定制设计了主管级别的办公桌，以满足用户的工作需求，同时办公桌还兼有小型会议室的功能。

内部庭院设计了一处水池，池中放置了一棵树。水，象征丰富充足；树，代表公司对环境的关注。公司老板是一个拳击手，还是一个空手道选手，将休闲活动区与内院相结合，老板可以在办公室内锻炼。

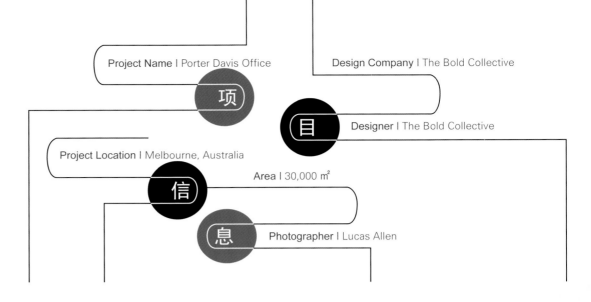

- Project Name | Porter Davis Office
- Design Company | The Bold Collective
- Designer | The Bold Collective
- Project Location | Melbourne, Australia
- Area | 30,000 m²
- Photographer | Lucas Allen

CORPORATE CULTURE

Porter Davis isn't just any building company. It started with a bunch of guys who wanted to make a difference in how customers were treated. That DNA lives on in our business. It's driven by our founder, expressed in our belief and reflected in our manifesto and comes to life in how we treat you. We take your dream seriously. Our job is to respect, protect and nurture your dream.

Porter Davis不仅仅是一家建筑公司,它由一群想要改变顾客待遇的人发展而来。现在,这个最初的"DNA"仍存在于公司的事业中,反映在公司的宣言中,并表达在员工的信仰中,在公司对待顾客的方式上得以实现。公司认真对待顾客的梦想,以尊重、保护、培育顾客的梦想为重要工作。

DESIGN CONCEPT

They engaged us with a clear brief. The main objectives are to accommodate more over 300 staffs within one floor plate at 720 Bourke Street and reflect their unique World Of Style design process within the workplace. Currently, their staff is widespread across seven locations throughout Melbourne and this co-location project aspired to build and strengthen culture and enable better work practices. An activity based workplace approach is adopted with a variety of work settings and a number of client facing meeting rooms.

Porter Davis believes that people's homes are the ultimate expression of who they are and have developed their World of Style design process and showroom to enable their clients to express themselves through their home styling and interior design, taking aesthetic cues from around the globe. Porter Davis engages us to make sure their new workspace reflected these ideals.

该办公室项目的主要目标是将300多名员工统一到位于Bourke大街720号建筑的同一个楼层,并且创造出一个能反映公司具有世界风格的独特设计理念。目前,他们的工作人员遍布墨尔本7个地区,这一共同办公项目的目的便是建立和加强公司文化,并令工作得以更好地实践。空间采用了以活动为基础的工作场所方法,并设置了各种工作环境和一些面向客户的会议室。

Porter Davis相信,人们的家庭住房能够最大限度地反映出他们是什么样的人。其公司从世界各地汲取审美灵感,开发出了属于自己的设计风格、工作流程以及样板间,让客户能够通过他们自己家庭的装修和室内设计,表达出内心的真实想法。Porter Davis让设计师确保他们的新工作空间反映了这些理想。

SPACE FEATURES

We work very closely with Porter Davis's in-house designers to realize their new, unique workplace project. It's a dynamic place to work and visit and provides an inspired and unified base that enables a thriving workplace culture.

The aesthetics for the project reference Porter Davis's World Of Style model, where clients are invited to participate in a style evaluation to assist in the design of their new home. The online survey takes clients through a number of images that are categorized in different styles based on destinations which range from New York to London to Bali to Milan. At the end of the survey, the results are collated, and the clients overall style is revealed. This style can be translated into different finishes, colors, materials and textures throughout their new home. For example, a Berlin home would have classic lines, aged timbers or concrete floor, large white walls, industrial undertone, mono tonal gray palette, factory style lights and the gritty feel to textures in fabrics. A Fifth Avenue New York home might feature dark timbers and black furniture, stone and luxurious velvet upholstery with chrome and glass highlights throughout. These more than 60 styles are reflected in Porter Davis' World Of Style showroom and form the cornerstone of the new Porter Davis workplace.

Stepping out from the lift at 720 Bourke Street, visitors and staff will experience a New York arrival and reception area. This includes a sleek reception desk with turned timber legs, cut crystal decorative lighting, a recycled timber herringbone strut ceiling and New York loft mullion detailing to the glazed meeting room partitions. Walking further into space, you will encounter a French Industrial inspired kitchen, open grid ceiling and industrial fixtures and fittings. The Melbourne area includes local street art, custom printed work tables and screens referencing iconic Melbourne locations and grid city planning. The Resort areas of Bali, Maldives and Portsea Beach House are strategically located in the quiet areas of the workplace and include a suspended swing and large comfortable lounges to encourage contemplation and quiet work. The London area includes a large Union Jack graphic. Also, within the workplace are both Nautical Hamptons and Classic Hamptons areas with a Nautical Booth with navy and white striped fabric, rope wrapped table legs, portholes and decorative oars.

设计团队与Porter Davis的室内设计师紧密合作，以实现他们独特的新办公室项目。它是一个充满活力的工作和参观场所，提供了一种充满灵感和统一的繁荣的职场文化。

该项目的审美参考Porter Davis的世界风格模型，Porter Davis被邀请参与风格评估，以协助设计他们的新家。在线调查请Poter Davis根据不同的风格，对多幅纽约、伦敦、巴厘岛和米兰的图片进行分类。在调查结束后，设计团队对调查结果进行了整理，揭示了Porter Davis的总体风格。这种风格可以被转换为不同的饰面、颜色、材料和纹理，呈现在他们的新家。例如：柏林的房子的风格会有古典的线条、旧的木地板或混凝土地板、白色的大墙壁、工业风的底色、单色调的灰色调色板、工业风的灯光和粗糙的织物纹理；纽约第五大道的房子的风格可能会采用深色的木材和黑色的家具、石头和带有铬合金和玻璃饰面的奢华天鹅绒。60多种风格在Porter Davis的世界风格陈列室里呈现，并构成了新的Porter Davis办公室的基石。

从Bourke大街720号电梯下来，游客和工作人员将体验到纽约式的接待区。这包括一个圆滑的接待室，转动的木腿、切割的水晶装饰照明、可循环利用的人字形木质天花板和纽约阁楼的竖框，详细展示了光滑的会议大厅。走进更深入的空间，人们将遇到一个灵感源于法国工业风的厨房、开放的网格天花板、工业装置和配件。墨尔本风格区包括当地的街道艺术、定制的印刷工作表和屏幕，参考自墨尔本标志性的地理位置和网格城市规划。巴厘岛、马尔代夫和波特西海滩别墅的度假区位于整个办公室的安静区域，里面有一个悬空的秋千和大型舒适的休息室，以鼓励人们进行思考和安静地工作。伦敦区有一大幅英国国旗。此外，工作场所内还有航海汉普顿和普通汉普顿地区的特色，海军蓝和白色条纹物、绳包桌腿、舷窗和装饰桨的航海风的隔间，令人不由地眼前一亮。

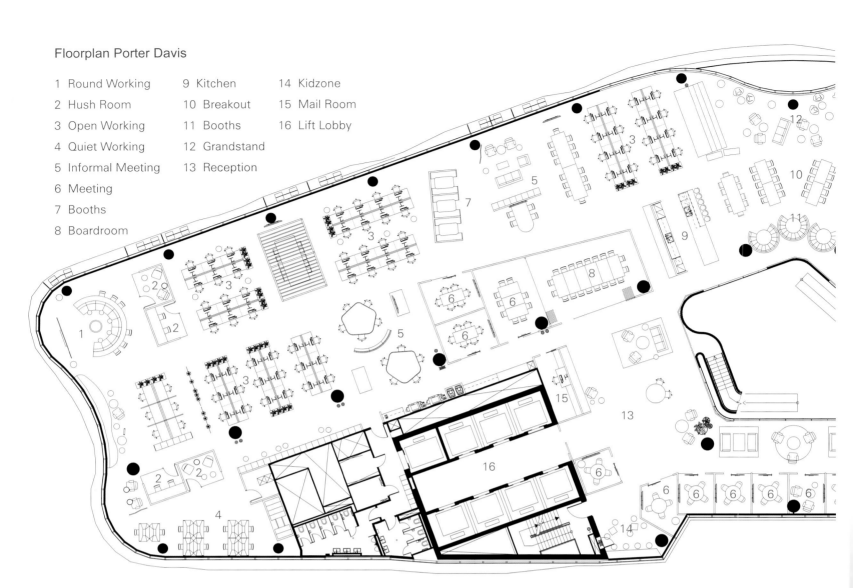

Floorplan Porter Davis

1. Round Working
2. Hush Room
3. Open Working
4. Quiet Working
5. Informal Meeting
6. Meeting
7. Booths
8. Boardroom
9. Kitchen
10. Breakout
11. Booths
12. Grandstand
13. Reception
14. Kidzone
15. Mail Room
16. Lift Lobby

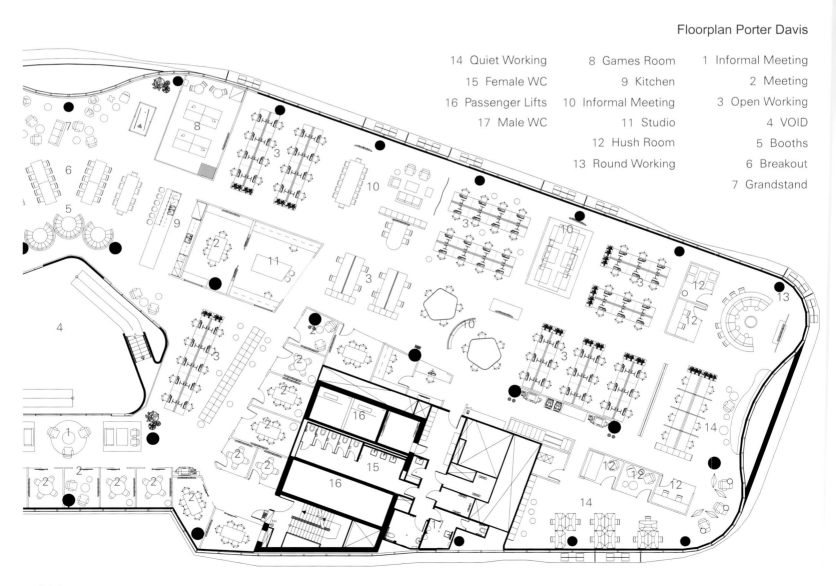

Floorplan Porter Davis

1. Informal Meeting
2. Meeting
3. Open Working
4. VOID
5. Booths
6. Breakout
7. Grandstand
8. Games Room
9. Kitchen
10. Informal Meeting
11. Studio
12. Hush Room
13. Round Working
14. Quiet Working
15. Female WC
16. Passenger Lifts
17. Male WC

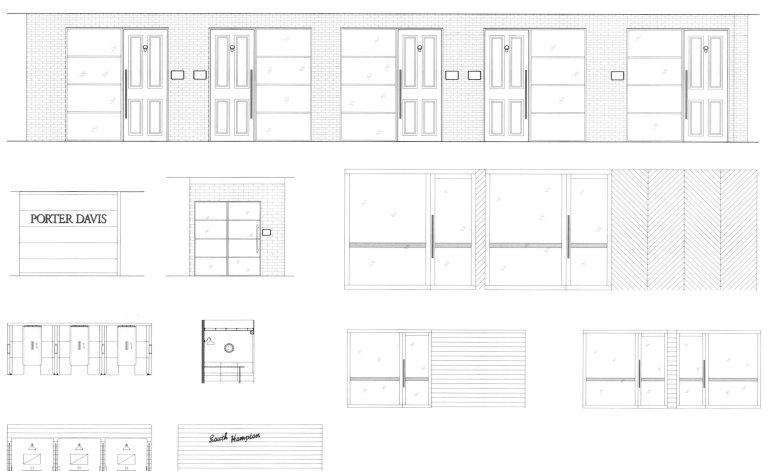

图书在版编目（CIP）数据

定制办公. Ⅳ, 诠释企业文化与办公空间的融合 / 深圳视界文化传播有限公司编. -- 北京：中国林业出版社，2018.4
　ISBN 978-7-5038-9560-9

　Ⅰ. ①定… Ⅱ. ①深… Ⅲ. ①办公室－室内装饰设计－作品集－世界－现代 Ⅳ. ① TU243

中国版本图书馆CIP数据核字（2018）第 099245 号

编委会成员名单
策划制作：深圳视界文化传播有限公司（www.dvip-sz.com）
总　策　划：万　晶
编　　　辑：杨珍琼
校　　　对：陈劳平　尹丽斯
翻　　　译：侯佳珍
装帧设计：叶一斌
联系电话：0755-82834960

中国林业出版社 · 建筑分社
策　　　划：纪　亮
责任编辑：纪　亮　王思源

出版：中国林业出版社
（100009 北京西城区德内大街刘海胡同 7 号）
http://lycb.forestry.gov.cn/
电话：（010）8314 3518
发行：中国林业出版社
印刷：深圳市国际彩印有限公司
版次：2018 年 4 月第 1 版
印次：2018 年 4 月第 1 次
开本：235mm×335mm，1/16
印张：20
字数：300 千字
定价：428.00 元（USD 86.00）